PLASTICS AND MICROPLASTICS IN ENVIRONMENT
AN INTRODUCTION

PLASTICS AND MICROPLASTICS IN ENVIRONMENT
AN INTRODUCTION

Carlos Andres Garcia Sanchez

Delve PUBLISHING

www.delvepublishing.com

Plastics and Microplastics in Environment: An introduction

Carlos Andres Garcia Sanchez

Delve Publishing
4164 Lakeshore Road
Burlington, ON L7L 1A4
Canada
www.delvepublishing.com

Email: orders@arclereducation.com

ISBN: 978-1-77956-354-5 (Hardcover)

ABOUT THE AUTHOR

Carlos Andrés García is an environmental specialist with over 14 years of experience in climate change, sustainability, and international cooperation. He has led and managed projects aimed at fostering resilience, poverty reduction, and environmental sustainability. Currently serving as a Climate Change Consultant, Carlos has designed and supervised numerous climate change adaptation projects, securing international funding from bodies such as the Nordic Fund, USAID, and the Green Climate Fund. Throughout his career, Carlos has written numerous scientific articles focused on conservation programs, climate development, and environmental governance. His publications have contributed significantly to advancing knowledge in areas such as climate resilience, marine conservation, and environmental governance. Carlos's professional journey includes his work as a senior consultant for international development banks, where he assessed climate-resilient solutions for sectors like ports, coastal infrastructure, and agriculture. His expertise extends to navigating international environmental standards, promoting Nature-Based Solutions (NbS), and managing large-scale conservation programs. He has played a pivotal role in enhancing the environmental governance of protected areas and supporting local communities in adaptation strategies. In addition to his work in project management, Carlos holds a Master's in Environmental Economics from the University of Western Brittany in France and a Bachelor's in Business Administration from Colombia. He is also a certified Climate Change Officer. His dedication to sustainable development, along with his extensive writing and research experience, makes him a key contributor to climate action efforts across Latin America and globally.

TABLE OF CONTENTS

List of Figures .... *xi*

Glossary of Terms .... *xiii*

Preface .... *xv*

Introduction .... *xvii*

Chapter 1 Introduction to Plastics **1**

Introduction 2

1.1. Pollution Due to Plastics: A Menace 3

1.2. Transportation Problems 5

1.3. Polymers 6

1.4. Polyethylenes 10

1.5. Research Trends 12

1.6. Catalysts and Other Parameters for Resin Synthesis 13

1.7. Sensitivity of Polymers 17

1.8. Phthalates – Health and Other Aspects 19

1.9. Addressing Emissions 20

1.10. Chlorofluorocarbon and Research 22

Chapter 2 Introduction to Microplastics **27**

Introduction 28

2.1. Microplastics 28

2.2. Effect of Microplastics 31

2.3. Surveys 33

2.4. Sampling Aspects 34

2.5. Soils 35

2.6. Samples and Their Processing 36

2.7. Techniques for Analyses 39

2.8. Processing 41

2.9. Reports From Research 43
2.10. MP, NP, and HDPE 44
2.11. Analytical Techniques of Microplastics 46

Chapter 3 **Microplastics and Their Characterization 51**
Introduction 52
3.1. Plastic Waste 53
3.2. Microplastics-More on Them 54
3.3. Occurrence and Effects 55
3.4. Associated Reactions 56
3.5. Microorganisms-Based Breakdown of Plastics 57
3.6. Consumption of Polymers and Effects 59
3.7. Research Findings 62
3.8. Sampling of Microplastics 63
3.9. Collection Details 65
3.10. Problems With Microplastics 66
3.11 Assessing Microplastics in Aquatic Environments 68
3.12. Identification Reports of Microplastics 71

Chapter 4 **Effect of Microplastics in Marine Ecosystems 75**
Introduction 76
4.1. Sustainability of Resins 76
4.2. Impact of Plastics 77
4.3. Pollution Due to Plastics and Additives 79
4.4. Plasticizers 80
4.5. Health Effects of Exposure 81
4.6. Pigments 83
4.7. Polyolefins 84
4.8. Packaging 86
4.9. Organotins 87
4.10. Codex Alimentarius 88
4.11. Chemicals 89
4.12. Bottle Materials 90
4.13. More on BPA 91
4.14. Research Findings 92

4.15. Antimony in Bottled Water 94
4.16. Floating Plastics 96

Chapter 5 Microplastics in Freshwater Systems 99
Introduction 100
5.1. Microplastics 100
5.2. Challenges to Studying Particles 101
5.3. Fragments From Degradation 102
5.4. Other Aspects of Polymers 103
5.5. Chemical Interactions 104
5.6. Particles as Pollutants 105
5.7. Risk Management 107
5.8. Detection of Materials and Associated Issues 109
5.9. Surface Interactions 110
5.10. Assessing the Risks 111
5.11. Other Assessments 113
5.12. Ecotoxicity of Nanomaterials 114

Chapter 6 Toxic Effects of Plastics 119
Introduction 120
6.1. Health Problems Due to BPA 121
6.2. BPA in Consumer Products-What Does Research Say? 121
6.3. Observations and Effects on Cancer 123
6.4. BPA – Does It Adversely Affect Human Fertility? 124
6.5. Endocrine Interference of BPA 125
6.6. Phthalates-Threat to Health 127
6.7. Other Toxic Materials 130
6.8. Consumption Patterns of These Contaminants 131
6.9. Presence in Household Articles 132
6.10. Engineered Nanomaterials 135
6.11. Other Synthetic Polymers 136

Chapter 7 Urban Management and Microplastics 139
Introduction 140
7.1. Distribution of Microplastics 141
7.2. Wastewater Treatment Plants 142

7.3. Activated Sludge Process 144
7.4. Sequencing Batch Reactor 145
7.5. Processing of Microplastics 146
7.6. Bacterial Dominance Patterns 147
7.7. Nitrogen Bacterial Dynamics 148
7.8. Changes In Microbes Due to Bioplastics 149
7.9. Research Findings 150
7.10. Inhibitory Effects of Polymers 152
7.11. Biological Treatment 153
7.12. Monitoring 154
7.13. Methane and Plastics 156
7.14. Microplastics Alter the Bacterial Abundance 158
7.15. Treating Microplastics in Wastewater 159
7.16. Mechanisms of Microplastic Damage 162

Chapter 8 **Human Health and Plastics 167**
Introduction 168
8.1. Toxicity to the Lungs 168
8.2. Skin Damage 169
8.3. Sensitivity 170
8.4. Carcinogens 171
8.5. Endocrine Disruptors 172
8.6. ECD Concerns 173
8.7. Proper Evaluation 175
8.8. Colorants-Concerns 176
Conclusion 178

Bibliography 179

Index 197

LIST OF FIGURES

Figure 1.1. Industrial revolution – timeline
Figure 1.2. Global ecosystem
Figure 1.3. Plastics damage the ecosystem
Figure 1.4. Thermoplastics
Figure 1.5. Polyvinyl chloride
Figure 1.6. Polystyrene
Figure 1.7. Polyethylenes
Figure 1.8. LDPE bags
Figure 1.9. Phthalates influence health
Figure 1.10. Chlorofluorocarbon
Figure 2.1. Plastic pollution in Antarctica
Figure 2.2. Microplastics
Figure 2.3. Storms transport microplastics
Figure 2.4. Remotely operated vehicles
Figure 2.5. Soils
Figure 2.6. Microplastics in the digestive tract
Figure 2.7. Bioaccumulation of microplastics
Figure 2.8. Optical microscope images of microplastics
Figure 2.9. Microplastics in the mussel *Mytilus galloprovincialis*
Figure 2.10. Vibrational spectra of MP
Figure 3.1. Biodegradable plastics
Figure 3.2. Great pacific garbage patch
Figure 3.3. Landfills
Figure 3.4. Persistent organic pollutants
Figure 3.5. Effects of microplastics on the soil
Figure 3.6. Bioaccumulation of microplastics
Figure 3.7. Neuston tow nets
Figure 3.8. Occurrence of each type of microplastics in the king mackerel (*S. guttatus*) from the lower Meghna estuary in Bangladesh examined under a stereomicroscope

Figure 4.1. Medical plastics
Figure 4.2. Computer housing
Figure 4.3. Food packaging
Figure 4.4. Codex Alimentarius
Figure 4.5. PET bottles
Figure 5.1. Risk management
Figure 6.1. BPA-its use in water storage
Figure 6.2. Phthalate contamination in food
Figure 6.3. Plastic trays
Figure 7.1. Wastewater treatment plants
Figure 8.1. Endocrine disrupting chemicals

GLOSSARY OF TERMS

Biodegradable Plastics: Plastics that can be broken down by microorganisms over time under specific environmental conditions, resulting in natural compounds such as water and carbon dioxide.

Bivalves: Aquatic mollusks that have a shell divided into two hinged parts, such as mussels and clams. They are filter feeders and are highly vulnerable to microplastic contamination.

Crystallinity: The degree to which a polymer exhibits a structured, ordered arrangement of its molecules. Higher crystallinity often leads to increased strength and density in materials such as polyethylene.

Fossil Fuels: Natural fuels such as coal, oil, or gas, formed from the remains of living organisms millions of years ago. Fossil fuels are commonly used in the production of plastics.

Microbeads: Tiny, solid plastic particles less than 5 mm in size, typically used in personal care products such as exfoliating scrubs. They are a significant source of primary microplastics in the environment.

Microfibers: Fine fibers, often released from synthetic textiles during washing, that contribute to microplastic pollution in aquatic ecosystems.

Microplastics: Small plastic particles less than 5 mm in size, often resulting from the breakdown of larger plastic items (secondary microplastics) or directly introduced into the environment (primary microplastics).

Monomer: A small molecular unit that can bond to other similar molecules to form a polymer. Common monomers in plastic production include ethylene and propylene.

Nanoplastics: Plastic particles smaller than 1 μm in size, which may have different properties and impacts compared to microplastics due to their minute size and higher surface area.

Phthalates: Chemical additives used to make plastics more flexible and durable, especially in products like PVC. Phthalates are controversial due to their potential health risks, including endocrine disruption.

Polyethylene (PE): A common thermoplastic polymer made from ethylene. There are different types of polyethylene, such as low-density polyethylene (LDPE) and high-density polyethylene (HDPE), used in packaging and containers.

Polypropylene (PP): A thermoplastic polymer used in a wide variety of applications, including packaging, textiles, and automotive parts. It is known for its versatility and resistance to chemical solvents.

Polyvinyl Chloride (PVC): A widely used plastic, often found in construction materials like pipes, window frames, and medical devices. PVC can be rigid or flexible depending on the additives used.

Primary Microplastics: Small plastic particles that are manufactured to be 5 mm or smaller, such as microbeads, pellets, and microfibers.

Secondary Microplastics: Microplastic particles that form through the breakdown of larger plastic debris due to environmental exposure such as sunlight, wave action, or chemical degradation.

Thermoplastics: A type of plastic that becomes moldable when heated and solidifies upon cooling. Thermoplastics, such as polyethylene and polystyrene, are commonly used in packaging and manufacturing.

Vinyl Chloride Monomer (VCM): A chemical compound used to produce polyvinyl chloride (PVC). Exposure to VCM is a concern due to its potential carcinogenic effects.

PREFACE

A vast range of academic and scientific disciplines, including polymer and material sciences, environmental toxicology and chemistry, biogeochemistry, economics, sociology, public health, decision sciences, physics, global environmental change, and mathematics, are all touched upon in the rapidly developing field of microplastic and plastic studies. Plastics are widely used, inexpensive to make, and versatile. They enable us to achieve technical breakthroughs that were unthinkable even a century ago, but they present a significant environmental issue for the future. As we become more aware of the effects of these materials, we are actively thinking of solutions to clean up the garbage/waste patches in our oceans. Several projects have been proposed that would gradually gather the waste and destroy it properly. Microplastics are microscopic, invisible pieces of plastic that we are less familiar with. Although invisible, their impact is probably just as strong. Since it is nearly impossible to remove plastic waste from water, raising awareness of microplastics and considering ways to stop their production is the only way to address the issue.

There are various methods for producing microplastics. Furthermore, we now know that certain face and body washes contain microbeads, which usually go down the drain when we take a shower. These tiny plastic particles manage to evade water treatment facilities and end up in oceans. Microplastics can also result from the disintegration of larger plastic objects that have been exposed to sunlight. These materials become brittle and discolored and are constantly being pounded by waves, which break them down into thousands of small plastic particles. Animals are at immediate risk due to these microplastics. They are consumed by a variety of aquatic species or pulled into their bodies by water currents as they pass through them in search of food. Particularly vulnerable are filter-feeding bivalves, such as mussels. Pollutant levels linked to microplastics increase significantly compared to the surrounding environment. Therefore, compared to waters devoid of these micropollutants, aquatic life in waters containing microplastics is likely to consume higher concentrations of pollutants.

This book *Plastics and Microplastics in the Environment: An Introduction* introduces the nature of plastics and microplastics and can be used by students and researchers in the field.

—Author

INTRODUCTION

Any kind of plastic pieces smaller than 5 mm are referred to as microplastics. They can enter natural ecosystems through a number of channels, such as clothes, cosmetics, and industrial activities. Primary microplastics are any plastic fragments or particles that, at the time of release into the environment, were no larger than 5.0 mm. These consist of plastic pellets, microbeads, and clothing-related microfibers. Alternatively, after entering the environment, bigger plastic items degrade due to natural weathering processes, producing secondary microplastics. Fishing nets, plastic bags, microwave containers, and tea bags are a few examples of secondary microplastic sources. It is known that both kinds are highly persistent in the environment. Microplastics have a high chance of being consumed and accumulating in the bodies and tissues of numerous species due to the slow rate of plastic degradation. Toxic substances from the ocean have the ability to biomagnify. These days, plastics and rubbers are two very distinct but equally vital materials that we use and depend on in our daily lives, both inside and outside our homes. Nonetheless, there is still debate concerning the use of some of them almost constantly, which has led to some misconceptions about their application. These misconceptions include scare tactics and false information regarding the material damage to human health, such as the notion that using polyvinyl chloride (PVC) in any way increases the risk of cancer. Nevertheless, it is now understood that the cause of these adverse effects is not PVC per se, but rather its monomer, vinyl chloride, and a few unique chemicals that are combined with it. Furthermore, PVC blood bags are actually used in medical applications.

Chapter **1**

INTRODUCTION TO PLASTICS

Contents

Introduction 2

1.1. Pollution Due to Plastics: A Menace 3

1.2. Transportation Problems 5

1.3. Polymers 6

1.4. Polyethylenes 10

1.5. Research Trends 12

1.6. Catalysts and Other Parameters for Resin Synthesis 13

1.7. Sensitivity of Polymers 17

1.8. Phthalates – Health and Other Aspects 19

1.9. Addressing Emissions 20

1.10. Chlorofluorocarbon and Research 22

INTRODUCTION

A confluence of circumstances has resulted in the species to have ever lived on Earth growing at a rate that is now unsustainable given the planet's resource base (Hagens, 2020). The majority of the food needed to sustain this huge population and their activities is produced either directly or indirectly by the solar insolation of Earth. This is mostly the solar energy that plants absorb during photosynthesis and then store and distribute throughout the food chain. Our modern lifestyle necessitates additional energy sources in addition to the energy found in food, mainly for transportation and industrial goods manufacturing. Humans have evolved technology to tap fossil fuel reserves, especially petroleum resources, to augment the small budget of day solar insolation, in contrast to other species that grow within a diurnal solar energy budget (Kalair et al., 2021). Since the Industrial Revolution (Figure 1.1), fossil fuel deposits that have been produced over millions of years by plants collecting and storing solar energy have been extremely beneficial to humanity.

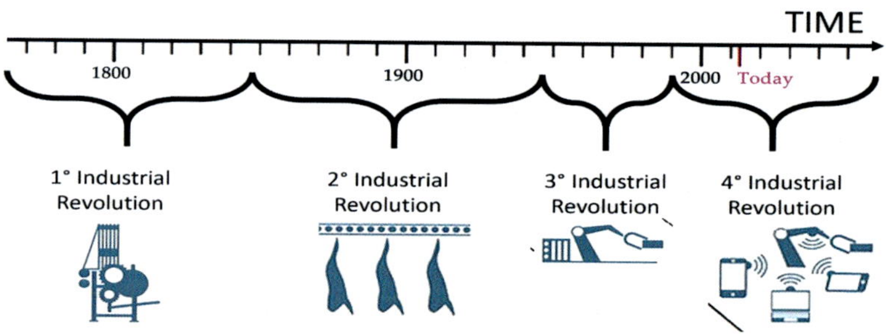

Figure 1.1. *Industrial revolution – timeline.*

Source: Zambon, I., Cecchini, M., Egidi, G., Saporito, M. G., & Colantoni, A. (2019). Revolution 4.0: Industry vs. Agriculture in a Future Development for SMEs. Processes, 7(1), Article 1. https://doi.org/10.3390/pr7010036, Creative Commons License.

The issue is that, in order to support our contemporary way of life, we must use these finite and rapidly depleting supplies far more quickly than the natural geochemical cycles can replenish them. Even worse, we not only waste these energy sources inefficiently by using them to produce the commodities and services that society needs, but we also harm the environment and generate a huge amount of waste in the process. The impact of pollution has expanded beyond its original association with diseases, birth deformities, and urban smog. The biosphere itself is starting to face threats from human activities. Acidification of water supplies, loss of biodiversity, and radiative forcing from greenhouse gases in the atmosphere are already beginning to affect the global ecosystem (Figure 1.2) (Singh and Singh, 2017). This is the first time in the planet's history that environmental destruction this severe and possibly irreversible has been caused by human activities.

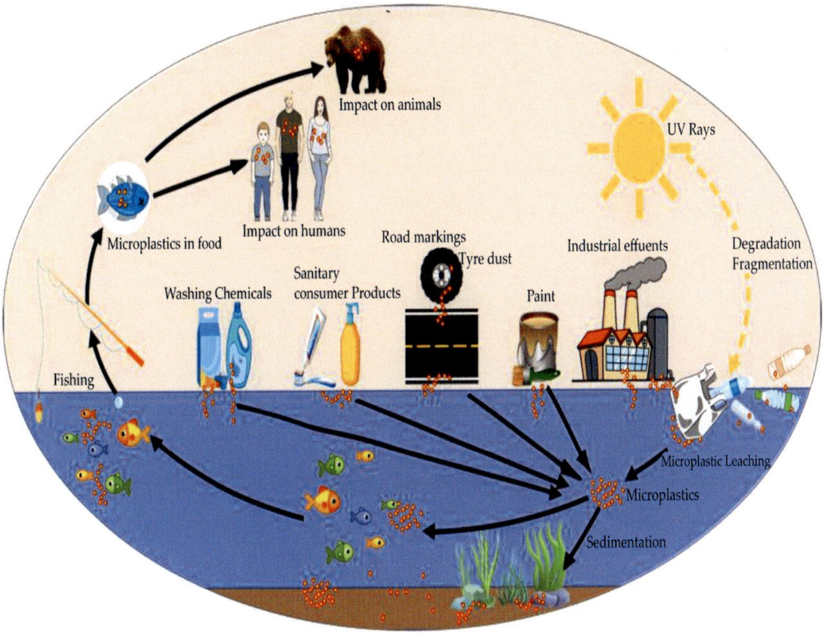

Figure 1.2. *Global ecosystem.*

Source: Ghosh, S., Sinha, J. K., Ghosh, S., Vashisth, K., Han, S., & Bhaskar, R. (2023). Microplastics as an Emerging Threat to the Global Environment and Human Health. Sustainability, 15(14), Article 14. https://doi.org/10.3390/su151410821, Creative Commons License.

1.1. POLLUTION DUE TO PLASTICS: A MENACE

Remarkably, compared to its potential consequences, this issue has received relatively little attention. The United States has, at most, made an inadequate commitment to studying the issue and its solutions. International progress on certain fronts, like the Montreal Protocol restricting the use of ozone-depleting chemicals, has demonstrated that working together to address global issues is feasible. However, there hasn't been much progress made on other important environmental issues like preserving biodiversity or lowering carbon dioxide emissions. With an annual growth rate of 4–5% between 1996 and 1999, the plastics business is doing well (Moore, 2008). Around a third of the world's resin production in 2000 came from North America, which accounted for about 135 million tones (Seyfullah et al., 2018). Major manufacturing regions that account for 20–23% of global production are also Western Europe and Asia. But in the future, as poorer nations increase their production and use of plastics, this situation is probably going to alter. Plastics are quickly catching up in price to the majority of conventional building materials, and their use will only grow in developing nations. Nowadays, industrialized nations use plastics in consumer goods at a comparatively higher rate. The consumption of plastics can reach 200 lb/person-year in wealthy countries like Western Europe or Japan, compared to less than 10 lb/person-year in the least developed nations (Andrady, 2003). However, developing countries, especially China, India, and several Latin American countries, are expected to have major growth in plastic consumption in the future. In densely populated countries, even a slight rise in the per capita consumption of plastics results in a significant increase in the amount of polymer used.

The developing world has also seen a rise in the use of disposable plastic packaging due to rising global wealth. The polymer-based industries, especially the low-tech processing operations are drawn to these nations by competitive wages, which promotes the use of plastics in developing nations even further. Over the past ten years, there has been little change in the significant difference in labor rates between workers in the plastics industry in the United States and nearby Mexico. Processes that include low labor costs, like fiber spinning, pipe or conduit extrusion, and injection molding, are being performed more often in developing nations. It should come as no surprise that manufacturing of plastics processing equipment is shifting from the West to emerging nations. Compared to Western Europe or the United States, the operational costs of the chemical industry are typically lower in developing nations due to the significantly laxer environmental standards. The potential contribution of plastics' manufacturing, usage, and disposal to environmental pollution is the main topic of discussion when it comes to plastics and the environment. Plastics must be manufactured using energy and raw materials, much like other man-made products like paper or glass. Similar to the manufacture of other materials, the production of plastics includes the use of fossil fuels, results in emissions, and generates waste that needs to be disposed of. In this sense, plastics are barely a special sort of material. The question at hand is not whether using and producing plastics causes a major disruption to the environment; rather, it is whether the environmental cost is eventually compensated by the benefits that plastics provide to society in terms of the economy, the environment, and other areas. The relevant issue to ask is if a different material with a lower environmental cost may provide the same level of benefits that plastics offer.

1.2. TRANSPORTATION PROBLEMS

The transportation sector uses more than half of all crude oil produced, and this application does cause substantial resource depletion and environmental damage (Curtis, 2009). However, there is currently no practical, affordable, and less polluting option to replace mass transit that runs on gasoline. The advantages of a well-functioning transportation system clearly outweigh the significant environmental impact that transportation has on society. The potential industrial emissions in the plastics sector, as well as any other industry, should be assessed in a manner commensurate with the very minute amount of valuable fossil fuel reserves used. Plastics (Figure 1.3) are a relatively new class of materials, having been around for barely 50 years, but they are already widely used in everything from airplane parts to medical implants. Concerns about the materials' environmental impact are starting to surface as consumer applications for plastics grow. One prominent environmental issue associated with plastics is the disposal of municipal solid waste. Given that packaging accounts for almost 30% of all plastic production, it is not shocking to discover that plastics are present in municipal solid waste and urban litter in considerable and increasing amounts. There has never been more consumer attention to the effects that solid waste has on the environment.

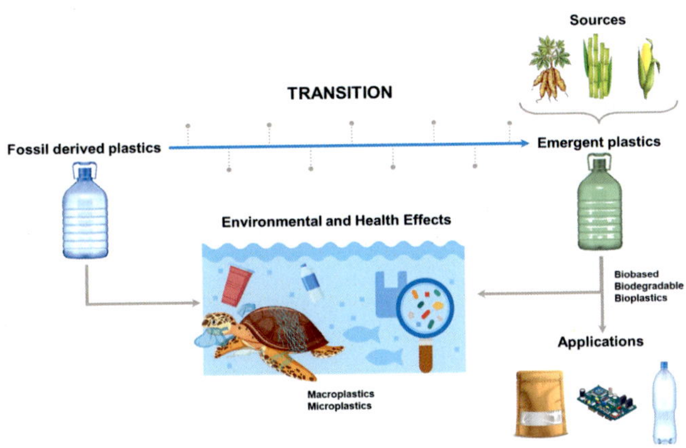

Figure 1.3. *Plastics damage the ecosystem.*

Source: Melchor-Martínez, E. M., Macías-Garbett, R., Alvarado-Ramírez, L., Araújo, R. G., Sosa-Hernández, J. E., Ramírez-Gamboa, D., Parra-Arroyo, L., Alvarez, A. G., Monteverde, R. P. B., Cazares, K. A. S., Reyes-Mayer, A., Yáñez Lino, M., Iqbal, H. M. N., & Parra-Saldívar, R. (2022). Towards a Circular Economy of Plastics: An Evaluation of the Systematic Transition to a New Generation of Bioplastics. Polymers, 14(6), Article 6. https://doi.org/10.3390/polym14061203, Creative Commons License.

1.3. POLYMERS

Leo Baekeland created the phenol formaldehyde resin, also known as Bakelite in Yonkers, New York (1907) (Seymour et al., 1989). The use of rubber and plastics has grown exponentially in just over 50 years, which is evidence of the cost-effectiveness, and adaptability of polymers as a class of materials. The peculiar molecular architecture of polymeric materials which consists of long, chain-like macromolecules gives polymers their remarkable properties. Although elastomers, or materials that resemble rubber, and plastics are both classified as polymers, conversations about environmental hazards have primarily focused on plastics due to their widespread use in construction and packaging. However, high-volume applications also use elastomers, especially in car tires where the product needs to be recycled or disposed of properly. Tires, which have a short lifespan, are always in great demand due to the increasing rise of automobiles. The environmental issues surrounding elastomers are unquestionably significant enough to merit a thorough investigation. These include the dangers of rubber chemicals to human health, the inhalation of rubber particles released into the atmosphere when tires break down while being used, and issues with disposing of used tires. A large number of the widely used thermoplastics (Figure 1.4) that are in use today were created after the 1930s, and several of them even surfaced after World War II.

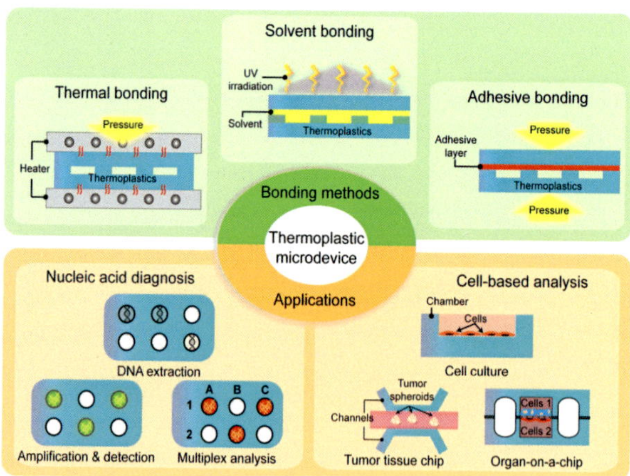

Figure 1.4. *Thermoplastics.*

Source: Trinh, K. T. L., Thai, D. A., & Lee, N. Y. (2022). Bonding Strategies for Thermoplastics Applicable for Bioanalysis and Diagnostics. Micromachines, 13(9), Article 9. https://doi.org/10.3390/mi13091503, Creative Commons License.

The vinyl polymers made of ethylene were among the first to be created. In 1928, I.G. Farben (Germany) and the Du Pont Company submitted the primary patents for the manufacture of vinyl chloride copolymers (Utracki, 1998). Although B.F. Goodrich discovered that the plasticized resin was somewhat helpful in waterproofing fabrics, the homopolymer Polyvinyl Chloride (PVC) (Figure 1.5) was already known but not thought to be particularly effective at the time.

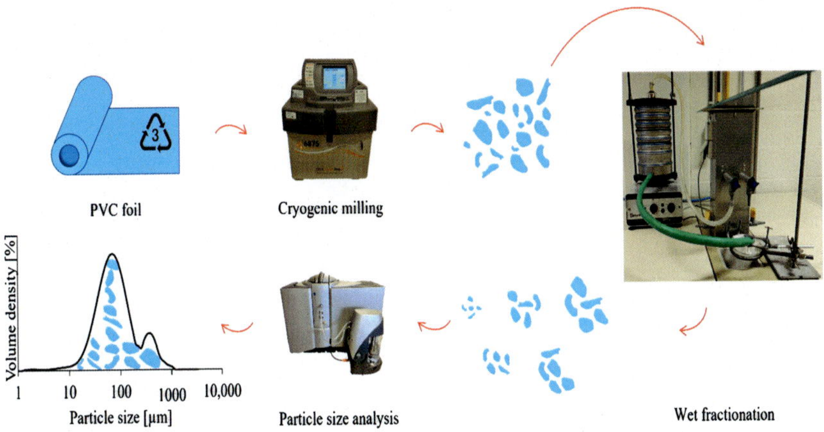

***Figure 1.5** Polyvinyl chloride.*

Source: Lievens, S., Vervoort, E., Poma, G., Covaci, A., & Van Der Borght, M. (2023). A Production and Fractionation Protocol for Polyvinyl Chloride Microplastics. Methods and Protocols, 6(1), Article 1. https://doi.org/10.3390/mps6010015, Creative Commons License.

Even before World War II, vinyl polymers and related plasticized compounds were being made on a commercial basis in the United States. However, the PVC that is used in construction today was developed after World War II and quickly increased in volume to the point where, by the early 1970s, vinyl resin demand was almost equal to that of polyethylene.

Imperial Chemical Industries (ICI) research laboratories discovered polyethylene, the plastic with the largest global usage, in 1933; nevertheless, the business did not file a patent for the relevant polymerization technique until 1936 (Hutley and Ouederni, 2016). For almost twenty years, until the development of low-pressure techniques for High-Density Polyethylene (HDPE) in 1954, Low-Density Polyethylene (LDPE) was only produced commercially by this high-pressure polymerization route. Even more recently, in the 1970s, came linear low-density copolymers of ethylene (LLDP), which are structurally and functionally intermediate between HDPE and LDPE. Another new class of polyethylene, based on metallocene catalysts, has been created in the last twenty years. The production of polypropylene began comparatively late in the 1950s, with the development of stereospecific Ziegler-Natta catalysts that produced high-molecular-weight propylene polymers (Jiang and He, 2014). Polyolefins are the collective name for polyethylene, polypropylene, and their common copolymers.

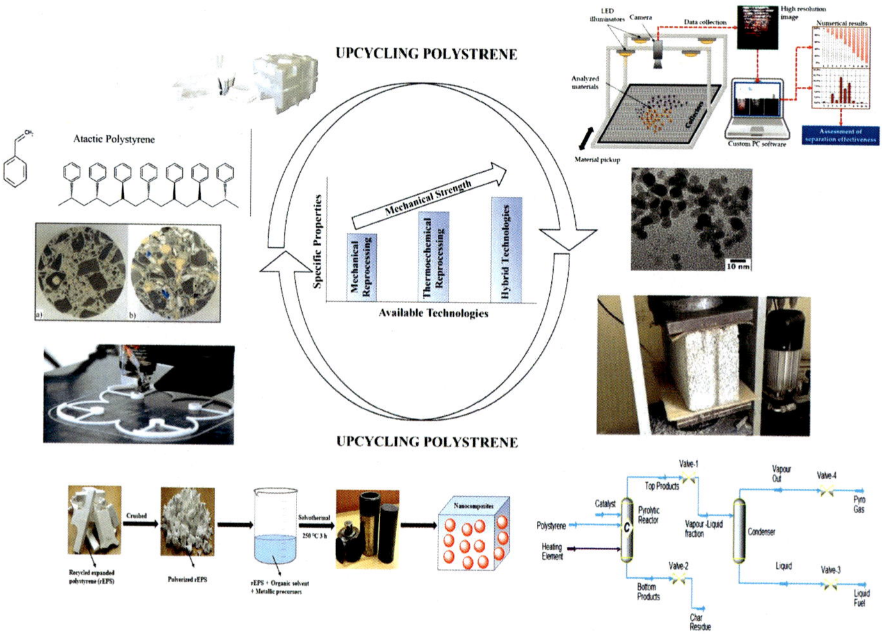

Figure 1.6. *Polystyrene.*

Source: Capricho, J. C., Prasad, K., Hameed, N., Nikzad, M., and Salim, N. (2022). Upcycling Polystyrene. Polymers, 14(22), Article 22. https://doi.org/10.3390/polym14225010, Creative Commons License.

Polystyrene (Figure 1.6) was created for the first time in 1930, and by 1934, facilities in Germany and the United States were manufacturing the resin for commercial use (Smith, 1988). Around the same time, ICI produced poly methyl methacrylate or PMMA. Nylon was widely utilized in daily items like women's stockings and toothbrushes, as well as tyre reinforcement and parachute cloth. In contrast, engineering thermoplastics, like polycarbonate, are relatively new, having been commercialized by General Electric Company in the mid-1958. Although other raw resources like coal or even biomass may also be utilized, petroleum and natural gas feedstock account for the majority of the millions of metric tonnes of polymer resins produced annually worldwide. Petroleum or coal tar is the only feedstock used in parts of the world where natural gas is scarce. Nowadays, petroleum accounts for about half of the polyolefins produced in the US; natural gas accounts for the remaining portion. To extract the lighter components, like gases, gasoline, and kerosene, the crude oil is distilled. The heavier components of crude oil from this distillation are catalytically converted into lighter, more usable components through the process of cracking. When crude oil enters a refinery, about 45% of it is transformed into gasoline. It is possible to directly polymerize or transform the ethylene that is produced when alkane gas mixtures or the naphtha portion break into usable monomers. These include vinyl acetate, vinyl chloride, and ethylene oxide, which can be used to create ethylene glycol. Similarly, the propylene fraction can be transformed into ethyl benzene and vinyl chloride.

Because of its abundance in benzene, toluene, and xylene, the catalytic reformate has a high aromatic percentage known as BTX, which serves as a vibrant source of raw materials for the production of aromatic polymers. These include benzene for the production of styrene and polystyrene, o-xylene for phthalic anhydride, and p-xylene for polyesters. When coal is utilized as the feed, it can be processed to produce water gas, which is a mixture of hydrogen and CO and is employed as a starting point for the synthesis of monomers. As an alternative, the monomers can alternatively be synthesized using acetylene that was obtained from the coal via the carbide method. Since the manufacture of plastics has increased rapidly in recent years, there is a strong correlation between population expansion and plastic output. Just fifty years ago, the world's resin production was negligible. Today, it exceeds 132 million metric tonnes per year (Kuczenski and Geyer, 2010). Extrapolating to a world population of approximately 10 billion by the year 2100, when population growth is predicted to stabilize, indicates an annual resin production of nearly 500 million tonnes worldwide (Scherbov, 2004; Geyer, 2020). Except for a temporary decline in 1995, the demand for commodity thermoplastics expanded steadily in the United States during the 1990s. Roughly 25% of the demand for thermoplastics in the United States comes from packaging applications, which are particularly interesting from an environmental perspective due to packaging in municipal waste and urban litter (Di et al., 2021). The uses of HDPE in bottles and containers, PP in fiber and closures, and PVC in plastics have had the fastest domestic growth rates in recent years. Due to their enormous populations, developing nations are becoming important major markets, even with relatively modest per capita incomes. Asian nations, especially China are positioned to play a significant role in the global plastics industry going forward. Asia processes less plastic than North America, which processes 34% of the total (Zhao and Schroeder, 2010). Asia processes about the same volume of plastic as Western Europe. China accounts for around half of the processing activity in Asia. It should come as no surprise that China leads the world in the production of machinery used in plastic processing. Furthermore, it is predicted that 20% of the manufacturing capacity needs to be replaced. This will require major technological advancements and increased capacity for the industry. The Pearl River Delta area of Hong Kong, which also produces processing machinery, has a thriving industrial sector. China, which ranks fifth in the world for resin production, uses very little of the resin produced there. Only plastics that are utilized widely enough and in sufficient quantities have a noticeable and significant impact on the environment. This is especially true for problems pertaining to solid waste, where polystyrene foam and polyolefin packaging materials have received a lot of attention. Moreover, the amount of environmental effects often rises with a material's global production volume.

1.4. POLYETHYLENES

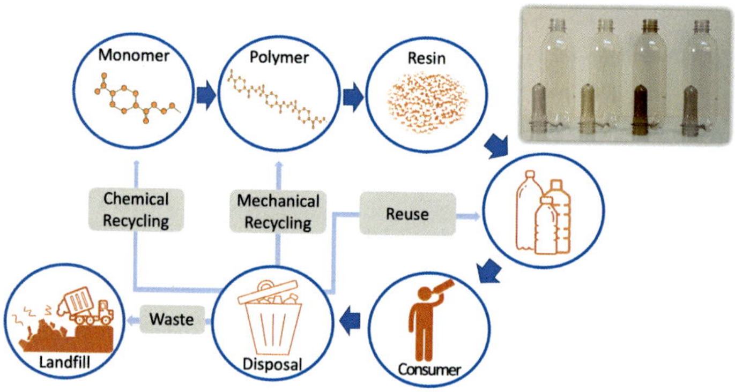

Figure 1.7. *Polyethylenes.*

Source: Benyathiar, P., Kumar, P., Carpenter, G., Brace, J., & Mishra, D. K. (2022). Polyethylene Terephthalate (PET) Bottle-to-Bottle Recycling for the Beverage Industry: A Review. Polymers, 14(12), Article 12. https://doi.org/10.3390/polym14122366, Creative Commons License.

Commercial polyethylenes (Figure 1.7) exhibit a wide range of average molecular weights, molecular weight distributions (polydispersity), and chain branching in the resin, depending on their copolymer composition and the polymerization method employed. These molecular characteristics regulate the material's degree of crystallinity as well as the macromolecules' capacity to pack tightly into a dense matrix. Polyethylenes have a fairly broad variety of physical properties due to their semi crystalline structure, which prevents them from exhibiting their theoretical density of 1.00 g/cm3 or the theoretically predicted melting point of roughly 135°C (Samuels, 1969). These, in particular the bulk density, are used to categorize the resins into three main groups. They are Low-Density Polyethylenes (LDPE), High-Density Polyethylenes (HDPE), and Linear low-density polyethylene (LLDPE). The simplest structure is found in high-density polyethylene, which is primarily composed of long, nearly unbranched polymer chains. Because these chains can easily align and pack, HDPE is the polyethylene with the highest degree of crystallinity. A degree of crystallinity as high as 70–95% can be obtained by having a molecular weight that is high enough and minimal chain branching, which results in a high density in the range of 0.941–0.965 g/cm^3 (Kiparissides et al., 1993). In contrast, the structure of low-density polyethylene exhibits excessive chain branching. There is typically both long and short-chain branching, which leads to a relatively lower material density of 0.910–0.930 g/cm^3 and only 40–60% crystallinity (Patel, 2017). The melting point of LDPE is between 110 and 115°C. By adding a minute quantity of an appropriate comonomer, the resin's crystallinity and melt temperature can be further decreased. A linear low-density polyethylene with a density range of 0.915–0.940 g/cm3 and a higher degree of crystallinity of 40–60% is obtained when the branches on the polymer chain are primarily short chains. The LLDPE resin was first introduced in 1968 and has since been widely used in packaging films, especially for items like garbage sacks and supermarket bags where high clarity is not a need. Short branches of LLDPE produce a very strong and durable material that may frequently replace LDPE packaging film with LLDPE that is only roughly one-third as thick. Future packaging applications will probably use more LLDPE due to its cost-effectiveness.

A new generation of LLDPE resins, known as Metallocene LLDPE resins, (mLLDPE) is already on the market (Knuuttila et al., 2004). These resins are based on metallocene catalysts. With the use of this novel catalyst system, LLDPE and Very Low-Density Polyethylene (VLDPE) resins with precisely regulated levels of short-chain branching and a remarkably narrow molecular weight dispersion can be produced. The mLLDPE resins' microstructure and mechanical characteristics are different from those of LLDPE produced with traditional Zieglar-Natta catalysts. The polydispersity of mLLDPE resins is getting close to the value of 2.0 as opposed to the usual values of 20–50 for LDPE and 4–15 for HDPE (Ronca, 2017). A resin with a smaller molecular weight distribution would typically have superior mechanical qualities, such as higher toughness or hardness and improved low-temperature impact strength. Nevertheless, the same characteristics also make it challenging to process and melt the resin. Introducing controlled amounts of chain branching along with low concentrations of specific comonomers can frequently relieve some of these challenges. One important factor influencing the resin's characteristics is the degree of chain branching in polyethylene. When the extent of chain branching grows, key material qualities like hardness or the initial tensile modulus fall almost linearly. Branch points and comonomers in the chain decrease the possibility of crystallite formation at those locations, which lowers the resin's overall density and crystallinity.

1.5. RESEARCH TRENDS

Current market research points to excellent prospects for polyolefins in general and polypropylene in particular. While HDPE is still able to successfully compete with vinyl in the pipe market, LLDPE is currently increasing at the fastest rate among the polyethylenes. Its high modulus and extremely low permeability are especially useful in applications involving bottles, containers, drums, and housewares. In contrast to the translucent LDPE goods, molded pieces and thick sheets made of the high-density variant of polyethylene have the highest rigidity. As a result, the latter grade finds a sizable market in films used for irrigation pipes, greenhouse films, and packaging including plastic-coated paperboard products. Approximately two-thirds of the LDPE market in the US is made up of these. Currently, the garbage-bag market is the biggest one, but due to the greater strength and resistance of LLDPEs to tearing and punctures, this application is seeing a market growth. In contrast, LDPE bags (Figure 1.8) are more affordable than paper bags and may be produced with handles.

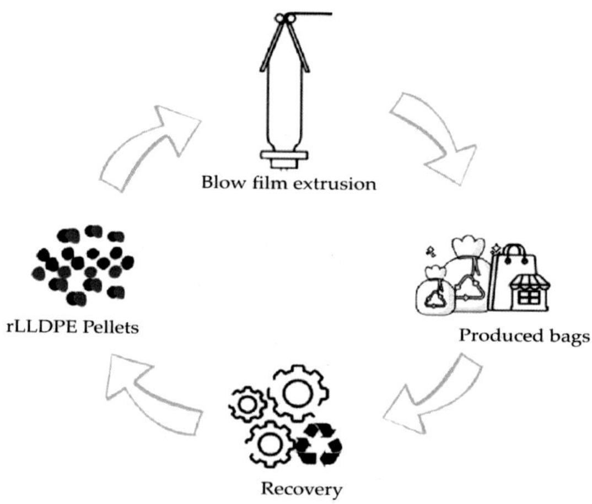

Figure 1.8. *LDPE bags.*

Source: Ballestar, R., Pradas, C., Carrillo-Navarrete, F., Cañavate, J., & Colom, X. (2022). Circular Economy Assessment in Recycling of LLDPE Bags According to European Resolution, Thermal and Structural Characterization. Polymers, 14(4), Article 4. https://doi.org/10.3390/polym14040754, Creative Commons License.

They are also tougher and thinner. Due to this, they have recently experienced growth in the paper bag market and that particular market sector, which also includes items like ice bags and clothing bags. Jugs for milk and juice make up the majority of blow-molded items, which make up over 40%

of the HDPE used in packaging. Film applications make up only around 7% of HDPE use, whereas injection-molded items such as food containers and crates consume only approximately 30% of the resin (Akelah, 2013). For the purpose of making polyethylene, ethylene is obtained by splitting different petroleum oil constituents, such as gas oil and the gasoline fraction, or from hydrocarbons like ethane. Although petroleum is still the most common source of the monomer nowadays, biomass can also be used to make it. Interestingly, molasses, a byproduct of the sugar cane industry, can be used to produce ethylene on a commercial scale by dehydrating ethanol.

The requirement for a separation phase to separate the resin product from the solution complicates these operations. The gas phase method, which can create both low- and high-density resins, is preferred by recent installations. Older plants can either generate low-density or high-density polyethylene; they are unable to create both types of polyethylene. Because of the reaction's high exothermic nature (22 kcal/mol), temperature control is crucial, particularly in autoclave reactors (Bajec et al., 2020). In general, there is a lot of long-chain branching in the product from chain transfer to polymer. When a different polymer chain is added to the mixture, the macroradical can also transfer, producing the longer branches that are normally present in high-pressure polyethylenes. In comparison to the resin produced in autoclave reactors, the product from the tubular reactors had a comparatively higher level of chain branching. Large, intricate plants were necessary for the high-pressure process, which necessitated strict process control.

1.6. CATALYSTS AND OTHER PARAMETERS FOR RESIN SYNTHESIS

Three different types of catalysts were developed: the Ziegler-type catalysts, which are typically produced by reacting alkyl aluminum compounds with titanium chloride; metal oxide catalyst systems, which are typically made of chromium oxide supported on a silicaceous carrier and were developed by Phillips Petroleum in the United States; and a different type of oxide catalyst that was developed by Standard Oil Company. By 1955, the first Ziegler catalyst-based facilities were operational in Germany, and in 1957, a Phillips catalyst-based facility opened its doors in Texas (Aftalion, 2012). The third catalyst system was adopted by the Japanese plastics industry at a factory that was inaugurated in 1961, although it progressed much more slowly. In 1960, Union Carbide and Du Pont launched the first LLDPE resin into the market (Karol, 1986). Organometallic catalysts, such as Ziegler–Natta catalysts, facilitated the polymerization process, which was conducted at a comparatively low pressure.

To regulate the density of the resin product, moderate concentrations of α-olefin comonomers are utilized. The density and stiffness of the resin are decreased and crystallinity is disrupted when appropriate comonomers are included into the polymer chains. Even with its better qualities, the molecular structure of LLDPE is far from ideal. The Molecular Weight (MW) distribution is wide and the comonomer is not evenly distributed throughout the various molecular weights. Another type of catalyst system, known as metallocenes, was created lately as a result of efforts to create a clearly defined restricted molecular weight distribution. A group IV metal atom is joined to two cyclopentadienyl groups and two alkyl halides or methyl alumoxane moieties to form the organometallic complex that serves as the catalyst. Gas-phase polymerization is a significant development in polyolefin production technology. Ethylene and any comonomers are continually fed into a fluidized-bed reactor in the Unipol (gas-phase) process at a pressure of roughly 0.7–2.0 MPa and a temperature of less than 100°C (Yates and Lettieri, 2016). The bed is immediately filled with the catalyst. The solubility of the resin product and issues related to handling and pumping viscous polymer solutions do not impede the gas-phase process. After the solid polyethylene is taken out of the reactor, any leftover monomer is cleaned out and put back. With minimal technical modification, the novel metallocene catalysts can be utilized by the gas-phase reactors.

Most significantly, the energy-intensive high-pressure polymerization process could be replaced with the substantially more energy-efficient low-pressure approach. The most versatile products in terms of resin density and polyethylene melt index are those produced by gas-phase methods. The gas-phase reaction is particularly interesting from an environmental perspective and has a number of advantages over traditional technology. The specific process type utilized in the production of polyethylenes determines the possibility of emissions. Apart from the typical risk of ethylene leakage in processing plants, there is also a requirement to properly monitor potential emissions of diluents and comonomers into the atmosphere. The manufacturing process does not appear to be a point source of these emissions, and the feed materials for the monomer and comonomer are generally harmless. However, there is a greater concern when it comes to the potential exposure of workers to some of the solvents and diluents because, in sufficient concentrations, they can target the Central Nervous System (CNS). The estimated emissions from the production of polyethylene were 1–18 for LDPE and roughly 13 for HDPE (Istrate et al., 2021). Monomer made up the majority of the fugitive emissions, with trace levels of lesser hydrocarbons like isobutane and ethane. During the production of polyethylene, certain particulate emissions also take place, and it is necessary to conduct an audit and analyze any potential health effects. The Volume Of Fugitive Emissions (VOC) from a facility has dropped due to advances in process efficiency, but as time goes on, the volume of resin produced increases, necessitating regular measurement and documentation of emissions. A known toxin utilized in the production process is the catalyst based on chromium (IV) oxide. When used in high quantities, it can harm the liver, impact the central nervous system, and possibly cause cancer.

The demand for polypropylenes has surged since the mid-1950s, when Ziegler-Natta catalysts were found, yielding high-molecular-weight resin. Because of their adaptability, the standard catalyst systems utilized in their production can be used to create homopolymers, copolymers, and random copolymers with different monomers. As a result of increased production, polypropylene is now sufficiently affordable to be employed in a range of low-value packaging applications, including tape, and containers. Currently, around 25% of the resin's global manufacturing capacity is accounted for by the United States. Propylene polymerization has also been carried out using the newly created metallocene catalysts. This new approach enables precise control over the resin's degree of stereoregularity and the copolymerization of propylene with a far larger range of comonomers. Although syndiotactic resins and other metallocene grades of polypropylene (mPP) are currently commercially accessible, they still make up a very minor portion of the total production. The production of polypropylene generally involves the low-pressure direct polymerization of propylene using Ziegler-Natta catalyst systems, which are usually titanium halides and aluminum alkyls with ether, ester, or silane activators. In traditional production, the process can be executed in liquid or slurry, or in fluidized-bed reactors. An isotactic index of 85–99 is typically the result of polymerization, which is commonly expressed as the percentage of insoluble in heptanes (Dumas and Hsu, 1989). Similar to polyethylenes, approximately 25% of polypropylene is produced as copolymers, particularly with ethylene. Flexibility and clarity are increased when ethylene units, even at as low as 0.3–5% by weight, are present because they break up the crystallinity in the polypropylene matrix. A random copolymer with 1–6% ethylene content produces a material with better flexibility, optical clarity, and a lower melting point. Heterophasic impact-grade copolymer is produced by adding more ethylene comonomer. A similar high-impact polymer can also be produced by mechanically combining polyethylene or ethylene-propylene rubber with polypropylene. Because the random copolymer in this instance is created with substantially higher quantities of ethylene comonomer, they can have an overall ethylene content of up to 15 weight percent.

There is a chance the low-cost resin produced by the ongoing demand for polypropylene will find its way into applications that currently employ other kinds of commodity thermoplastics. It is unlikely, nevertheless, to replace polyethylene unless it is needed for high-temperature applications, in which case isotactic Polypropylene (iPP) and its copolymers could be useful. The likelihood of such a substitute with rigid PVC materials seems feasible, with the exception of situations where the material must be regularly exposed outside. Although iPP's mechanical qualities are equivalent to those of PVC, its relatively low weathering resistance prevents it from being used outdoors. Nonetheless, such substitutes and even the replacement of some costly engineering thermoplastics with polypropylenes may become possible with the advancing technology of polypropylene and improved compounding procedures. When ethylene is created, steam cracking of hydrocarbons or catalytic cracking of petroleum fractions yields propylene monomers. Stirred reactors are used in conventional liquid phase and slurry processes, together with a diluent such as naphtha, hexane, or heptane. The reaction usually occurs between 60 and 80°C and between 0.5 and 1.5 MPa. The end result is a solid polypropylene suspension in a liquid phase. Centrifugation is one method that can be used to separate the resin, after which it must be cleaned of any remaining diluent and dried. Since the first polypropylene manufacturing facility was operational in 1957, numerous modifications have been made to the process. The Phillips and Novolon gas phase processes for polymerizing liquid propylene were introduced in the 1960s. Although the lack of diluents in these procedures was an advantage, they usually resulted in worse catalyst performance and certain restrictions on the stereoregularity of resins. Maximum yield and stereoregularity might be attained in 1975 with the advent of enhanced third-generation catalysts, which facilitated the reaction at the same temperature but at a slightly higher pressure of 2.5–3.5 MPa. The liquid monomer systems in the new gas-phase reactors could be utilized with these catalysts, which were introduced by Montedison and Mitsui.

Instead of using ethylene monomer feed, polymerization-grade propylene (99.8%) is utilized. Additionally, ethylene is typically employed as a suitable comonomer. The majority of the benefits of gas-phase methods mentioned for polyethylene manufacture also hold for polypropylene synthesis. Injection molding is used to transform more than one-third of the polypropylene produced in the US into beneficial goods. For this purpose, a broad variety of resins with Melt Flow Indexes (MFI) ranging from 2 to >70 g/10 min are available (Wang et al., 2016). Most polypropylene used in North American markets is spun into fibers or injection molded into goods for use in various textile applications. The latter comprises woven polypropylene bags that are cut from aligned sheets and are used to package agricultural goods. Bottles, jars, crates, closures, and containers are among the frequently manufactured items. Approximately 10% of the polypropylene gets extruded into a film. For low-temperature applications, copolymers are favored over homopolymers. Food packaging uses polypropylene non-oriented packaging films as well as Biaxially Oriented Polypropylene film (BOPP). The former, typically with a surface coating, is utilized as a barrier film. General-purpose uses for non-oriented films include bandages, hygiene items, diaper linings, and clothing bags. Bottles and other containers are made with the popular process of blow molding polypropylene.

1.7. SENSITIVITY OF POLYMERS

Because the polymer is prone to heat and photodegradation, light stabilizers must be added to the resin in goods meant for outdoor usage. These formulations usually comprise fillers, lubricants, and other additives like a thermal stabilizer package to preserve the resin during processing. The stiff PVC materials, also known as unplasticized polyvinyl chloride or uPVC, or compounds without plasticizers are widely used in building items like window frames, pipes, and fittings. An opacifier, typically rutile titania, that efficiently absorbs harmful ultraviolet (UV-B) radiation is added to unplasticized PVC compositions meant for outdoor usage in order to shield the surface from UV-induced deterioration. In improperly protected formulations, light-induced dehydrochlorination and, to a lesser extent, oxidation processes are caused by the solar ultraviolet component, specifically the UV-B radiation (290–315 nm). This results in the release of titania pigment from the binder at the material's exposed surface, uneven yellowish discoloration of white vinyl surfaces, and, after extended exposure, a reduction in the product's impact strength. PVC is mostly used in the building industry in the United States; residential siding comes in second place with almost half of all resin sales going toward pipe, fittings, and conduit. By adding plasticizers like organic phthalates to the mixture, PVC resin can also be transformed into a flexible, soft, and rubbery substance. PVC that has been plasticized, Prefabricated PVC or pPVC, finds extensive application in the production of packaging film, roofing membranes, belting, hoses, and cable coverings. Calendering is used with pPVC to create films and sheets. The resin is also utilized in many household products and as a covering for paper and cloth. A tiny quantity of the plasticized film is used in packaging, for example, in meat wraps where the Food and Drug Administration (FDA) has given it approval for use in food-contact situations. PVC is the most adaptable thermoplastic among the commodity resins, both in terms of processing ease and application scope.

Although the process of polymerization was understood as early as the 1930s, bulk polymerization of high-grade PVC did not become possible until the 1960s (Nakajima et al., 2017). 1,2-dichloroethylene, which is produced by reacting ethylene with chlorine yields Vinyl Chloride Monomer (VCM). By oxychlorinating ethylene in the presence of Cu (II) chloride catalyst, the hydrogen chloride by-product created during this dehydrochlorination process is utilized to create an additional 1,2-dichloroethylene. The primary industrial user of chlorine is the PVC industry, which uses over 42% of the gas produced. More than 90% of the resin is now produced via suspension polymerization of VCM, which is triggered by free radicals. The reactor used for the monomers emulsion polymerization is similar to the one used for suspension polymerization, with the exception that vacuum is typically utilized for stripping. VCM polymerization is an exothermic reaction (+ 410 cal/g), hence in large-scale manufacturing, removing heat from the system is crucial (Gao et al., 2013). It is crucial to regulate the reaction's temperature since it affects the resin product's polydispersity and average molecular weight. In suspension polymerization, a monomer-soluble initiator is used to disperse the vinyl chloride monomer in water using a protective colloid or a surfactant to control the final particle size. As dispersion agents in the reaction mixture, gelatin, soaps, glycols, and pentaerythritol, or their combinations, can be employed. Usually, the polymerization process is conducted in a glass-lined reactor with regulated agitation, between 50 and 75°C, and a pressure of roughly 0.7 MPa. Typically, the reaction vessel is kept oxygen-free to avoid interfering with the free-radical polymerization reaction.

Extra care is taken throughout production to guarantee that personnel are shielded from monomer vapor. Relief valve emissions from monomer storage tanks are regulated by overpressure protection systems to stop monomer emissions into the atmosphere. After stripping, the resin product is cleaned and includes as much as 3% by weight of residual monomer when it exits the reactor. Usually, steam stripping is done in a continuous stripping column where resin slurry runs from the top of the column. To enable quicker stripping, the tower is kept at a temperature higher than the resin's glass transition temperature (Tg). Only parts per million of the monomers are present in the resin that exits the column, and the majority of this is likewise eliminated when the resin is subsequently dried in rotary driers. The residual monomer concentration of commercial resin is less than 1 ppm following this thorough cleaning procedure. Phthalates are added to soft, flexible PVC compositions at a weight percentage of 10–50 weight percent (Earls et al., 2003). Commonly used phthalates include Diisononyl Phthalate (DINP) and Di-2-Ethylhexyl Phthalate (DEPH). These products include vinyl sheets and flexible roofing membranes. Since these plasticizers tend to move slowly to the product's surface, they may come into contact with people or the environment. In fact, common plasticizers are widely thought to be consumed often with food and are present in the environment at low concentrations in most parts of the world. These popular plasticizers have low acute and chronic toxicity. Numerous human diseases have been linked to this low-level consumption of plasticizers, based on the biological effects that massive dosages of the plasticizer in animals have shown. In 1999, the European Commission outlawed the use of DEPH in toys intended for younger children (under three years old), while also promoting voluntary limitations on the plasticizer's use by the producers.

1.8. PHTHALATES – HEALTH AND OTHER ASPECTS

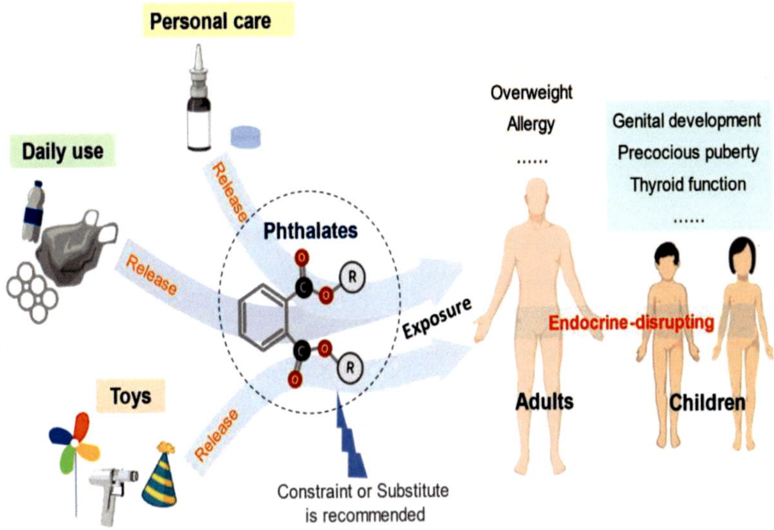

Figure 1.9. *Phthalates influence health.*

Source: Wang, Y., & Qian, H. (2021). Phthalates and Their Impacts on Human Health. Healthcare, 9(5), Article 5. https://doi.org/10.3390/healthcare9050603, Creative Commons License.

Phthalates (Figure 1.9) are not genotoxic. Though it is well known that they induce liver cancer in lab rats, it is not anticipated that this specific carcinogenic mechanism would work in people. There is no proof that DEHP causes cancer in human populations. PVC is used in hemodialysis equipment, blood bags, and infusion systems, among other medical devices. People who use these devices may absorb comparatively larger quantities of the dissolved plasticizer into their bodies. As a result, the toxicity of common plasticizers is a significant issue with health consequences. However, the general consensus among experts in the US is that plasticized PVC is safe to use in toys, medical applications, and other fields. Endocrine disruption by chemicals is another, more recent health problem, and plasticizers fall into this category of relevant chemical agents. It is unknown how PVC plasticizers contribute to the bioavailable environmental pool of endocrine disruptors, and it is yet unknown how relevant this is to human health. Lead, tin, barium, zinc, and cadmium compounds are among the frequently employed photothermal stabilizers in PVC (Ambrogi et al., 2017). It is commonly known that all lead compounds are toxic and that they can cause cancer, renal damage, infertility, and neurological problems in children. Morehazardous are cadmium and organotin chemicals, which have an impact on the kidney and nervous system. There is a significant risk that these metal compounds will leak out of PVC and into the environment in landfill settings. However, these chemicals are mostly trapped in the plastic matrix seen in landfills and potable water systems that use PVC pipes. It is doubtful that there will be any appreciable rates of chemical migration, unlike in plasticized systems. Therefore, it is not predicted that heavy metals will be released in any appreciable amounts into sewage ecosystems, landfill leachate, or water distribution networks.

However, this is not the case for PVC-related fires or for the unregulated burning of PVC waste. Due to its high halogen content, PVC is one of the least flammable common polymers in terms of ignition time; nonetheless, significant amounts of the metal compounds can be released into the atmosphere, making the plastic less flammable in a fire. Some members of the class of chemicals known as dioxins and furans, Polychlorinated Dibenzo-P-Dioxins (PCDD) and Polychlorinated Dibenzofurans (PCDF) are extremely harmful environmental contaminants. They may affect immunological and endocrine systems and are suspected human carcinogens. The main hazardous ingredient in Agent Orange was dioxin, which was also discharged in the Love Canal in Niagara Falls, New York, and Seveso, Italy. While some of these compounds are naturally occurring, others are man-made and are typically created when things burn. Nonetheless, the levels of PCDD/PCDF in the environment and in human milk have been steadily declining as a result of advancements in incineration techniques and emissions reductions. The majority of dioxins enter the body through diet (meat and dairy products). Nevertheless, it is unclear to what degree these substances are harmful to humans. Dioxin susceptibility varies greatly amongst test mammalian species.

1.9. ADDRESSING EMISSIONS

The primary cause of these harmful emissions is the burning of mixed garbage, which includes waste that contains plastics. However, it is unclear if altering the proportion of PVC waste in a waste stream will also alter the emissions of dioxin. Most people agree that burning at a high temperature in incinerators with suitable construction would reduce any dioxin emissions that may arise from the operation. For mixed waste streams, incineration combined with energy recovery can be a useful waste management technique. Given that waste streams usually contain a PVC component, it is critical to accurately measure and definitively determine its role in the production of dioxin during large-scale incineration. Because of the purity of the resin granules, general-purpose polystyrene is also known as crystal polystyrene. It is a hard, glassy substance that is clear and has a bulk density of 1.05 g/cm^3. The resin's moldability, minimal water absorbency, and good color range contributed to its popularity as a general-purpose resin. These favorable physical attributes were also important. Its brittleness, which restricted the items that the resin could be utilized in, was quickly remedied when high-impact hardened grades of rubber that contained polystyrene became accessible. There are several grades of the resin available, including general-purpose, high-impact, high-molecular-weight, high-heat, and easy flow. Mineral oil or another lubricant is added at a rate of 1–4% to the medium and easy flow grades to provide greater flow rates of approximately 7.5 and 16 g/10 min, respectively (Sikdar et al., 2020). High-heat resins are high-molecular-weight resins with melt flow rates of 1.6 g/10 min. Impact-grade resin, which is frequently used in injection molding of consumer goods, supplies roughly half of the market for polystyrene. Commercially accessible copolymers include Acrylonitrile–Butadiene–Styrene (ABS), Styrene–Acrylonitrile Copolymers (SAN), and Styrene–Maleic Anhydride Copolymers (SMA).

The Dow Chemical Company introduced a grade of Polystyrene (PS) in the United States in 1937, shortly after the Farben Company (Germany) began producing PS on a commercial scale for the first time (Boyer, 1981). The primary process for producing styrene monomer is the dehydrogenation of ethylbenzene, which is created by combining ethylene and benzene in a Friedel–Crafts reaction with the aid of an aluminum chloride catalyst system. With this method, yields above 98% are not uncommon. The dehydrogenation is produced via a thermal cracking reaction that occurs at 630°C with the aid of a catalyst, which is often a combination of Fe_2O_3, Cr_2O3, and K_2CO_3. Vacuum distillation is used to remove the styrene from the product mix, which also includes unreacted ethylbenzene and other contaminants. Because the monomer readily autopolymerizes into a hard solid, polymerization during storage is prevented by adding a small amount of a free-radical reaction inhibitor. Union Carbide invented a technique that oxidizes ethylbenzene to produce a negligible quantity of styrene. Propylene and the ethylbenzene hydroperoxide produced by oxidation react to produce 2-phenyl ethanol and propylene oxide. Styrene is produced by dehydrating the latter chemical. Although emulsion or bulk polymerization can also be employed for this purpose, a procedure with a free-radical initiator is the primary method used in the commercial production of polystyrene. The solvent, which is usually used at a concentration of 2–30%, regulates the viscosity of the mixture. Butadiene rubber is added to high-impact polymer during polymerization to alter it for usage in injection molding and extrusion. By using a devolatilization extruder or flash evaporation, the solvent and leftover monomer in the crude resin are eliminated. Extruded foam sheets and other foam products are made in large quantities using expandable polystyrene.

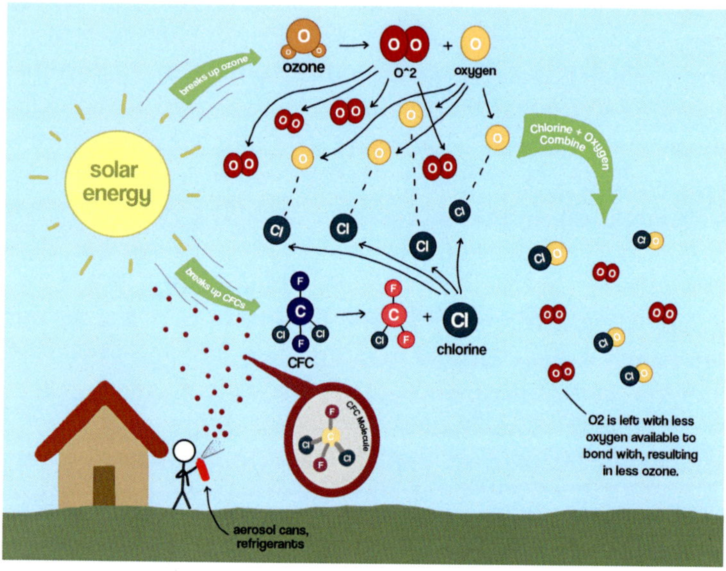

Figure 1.10. *Chlorofluorocarbon.*

Source: https://en.wikipedia.org/wiki/Chlorofluorocarbon#/media/File:CFCs_&_Ozone.jpg, Creative Commons License.

1.10. CHLOROFLUOROCARBON AND RESEARCH

A Chlorofluorocarbon (CFC) blowing agent, (Figure 1.10) which is currently restricted due to its potential to deplete stratospheric ozone, or a hydrocarbon blowing agent, such as pentane at a level of 5–8% by weight, may be used(Reimann et al., 2018). The U.S. industry phased out the use of ozone-depleting CFCs in expanded foams in the 1980s. A bicarbonate/citric acid system may be utilized as the nucleating agent with hydrocarbon blowing agents. The type and quantity of blowing agent used affects the foam's density and, consequently, its thermal conductivity, two factors that are crucial for construction applications. It is possible to achieve densities as low as 1 lb/ft^3 (16 kg/m^3), which translates to a grade of foam that is 97% by volume of air or blowing agent (Meister and Cummings, 2003; Niessner and Gausepohl, 2003). Usually, the bead is expanded in two steps using steam or hot air: first, it is preexpanded, and then it expands even more inside a mold, adhering to the beads. The manufacture of polystyrene foam cups, containers, and packaging materials uses molded expanded resin. Extrusion can be used to create foam sheets with a density range of 2–10 lb at 1 lb/ft^3 (32–160 kg/m^3) (Mhaske et al., 2023). Combining the blowing agent and resin in a preextruder allows for the extrusion of the mixture via a die to form sheets. These sheets could be thermoformed into serving trays or other consumer goods. Low-density polystyrene foams are an affordable way to achieve excellent insulation and bioinert containment. These benefits are demonstrated by the widespread usage of it in hot beverage cups and food containers, especially in the fast-food sector. Because of this, the product offers excellent value considering the very little quantities of fossil fuel resources required for its manufacture. Products made of expanded polystyrene foam serve as an excellent illustration of the significance of informing consumers about plastics' environmental implications. Unfortunately, the use of polystyrene foam in the packaging industry has been restricted by false beliefs over the usage of CFCs in the foams, the challenges associated with the practical recycling of postconsumer products, and exaggerated doubts about their disposal in landfills. The development emphasizes how crucial it is for the plastics industry to take aggressive steps in educating consumers about the environmental qualities of its products.

Polyester film with a biaxial orientation is used in thermoformed sheets for frozen meal trays that can be reheated in a regular oven or microwave, as well as in food packaging applications like boil-in-bag meals. PET films have excellent insulating qualities and are also utilized in electrical equipment (MacDonald, 2015). However, the most well-known item manufactured from aromatic polyester is the blow molded soda bottle, whose demand has increased dramatically in recent years. This use, which consumes 70–80% of the resin, takes advantage of the resin's superior carbon dioxide barrier qualities. Additionally, the resin must be treated in a way that results in an amorphous (transparent) polymer with strong gas barrier qualities. The grade of resin used in bottles has an average molecular weight of roughly 80,000. Terephthalic acid and ethylene glycol are used to create the condensation polymer known as polyethylene terephthalate. The process of oxidizing p-xylene, a byproduct of the catalytic reforming of naphtha, yields the acid or its dimethyl ester. The equivalent cyclic oxide is used to extract the glycol from ethane. Since pure terephthalic acid has been readily available since the 1960s, commercial polyester manufacturing has relied on the constant direct esterification of the acid. The processing sector has access to thermoplastic resins in the form of resin pellets. Different plastics industry segments may be involved in the conversion of the raw material into usable goods. The processor may alternatively perform the compounding in an internal facility.

To give the finished product particular qualities, a range of chemical additives must be thoroughly blended with the resin basic material. In the plastics business, additives are employed extensively in almost every kind of plastic product. Without the addition of additives, typical polymers could not be used in consumer goods. For example, vinyl polymers, especially PVC, are easily thermally and photodegraded; without stabilizer additions, the resin would not be protected during thermal processing and usage, and no viable products could be manufactured from it. Compounding is the process of choosing the right set of additives required by a particular product and combining them with the resin in the right amount. These can be added to the virgin resin as a concentrate that has already been disseminated in the same or a compatible resin in order to guarantee proper mixing or dispersion of the additive in plastic. It is also possible to put some additives straight into the resin mixture. In either case, the resin and additive mixture are mixed by feeding them through a mixing screw in an extruder at a temperature high enough to melt the thermoplastic. It is carefully monitored to ensure that neither the plastic nor the additional components undergo chemical breakdown as a result of overheating the mixture. After the resin has been compounded and the additives have been evenly distributed throughout, it is repalletized, chilled, dried, and stored for further processing. The last but one stage, known as processing, is what turns the compounded material into a functional plastic product. Essentially, the compounded resin must be heated to a temperature that facilitates the easy handling of the fluidized plastic, or melt, and melted into a liquid. In order to form the material into the desired forms, this melt is poured into molds and swiftly cooled to produce the product. Usually, the product needs to be slightly finished before it can be sold to customers. The various procedures followed and the machinery utilized depend on the kind of product being produced.

Extrusion, in which the plastic is heated in a metal tube, is the most important processing method for typical thermoplastics. An appropriately designed dye can be used to manufacture tubular products such as plastic drinking straws, plastic rods, plastic tubes, coatings on electrical wire, and fibers for textile applications. The design of an extruder is similar to that of a toothpaste tube. An Archimedean screw is employed to apply sufficient pressure to push the viscous melt through the tiny dye opening. Although external heating is also supplied, the mechanical shearing action of the screw provides the majority of the heat required to melt the resin. The resin is transported by the screw into a heated die from the input over a lengthy tunnel with multiple heating zones. Before the resin enters the mold, it goes through a section of the screw that guarantees additional mixing and consolidates the melt, eliminating any air pockets or empty spaces. The geometric properties of the extruded product are determined by the dye design. An accurately drilled hole or slit that produces a rod or ribbon product is the most basic type of die. Pipes and tubes are made with a little more complex design. A specialized dye is utilized to cover electrical conductors with thermoplastic polymers. The molten polymer introduced from the top of the die comes into touch with the conductor as it is pulled through the cylindrical die. Complex dyes are used in the extrusion of complex profiles, such as those seen in plastic door and window frames (Dogu et al., 2021).

When a pipe is cut to size, downstream equipment handles the product that emerges from the die, collecting and essentially cooling it for storage. The actual tools needed for the job vary depending on the kind of product produced. One of the most often used processing techniques in the plastics sector is injection molding. Injection-molding machines accounted for more than half of the processing machinery manufactured in recent years. The apparatus is primarily made to melt resin, inject the melt into a cavity mold, pack material into the mold under intense pressure, allow the mold to cool to produce a solid product, and then eject the product for further finishing. Unlike extruders, it uses a mold rather than a die, and packing the melt into the mold requires a lot of force. A machine is usually categorized based on its clamping force, which is based on the size of the product that needs to be made. Generally speaking, the machine consists of the systems like hydraulic, plasticating and injection, mold system, and clamping system. The equipment is powered by the hydraulic system, which is very useful for opening and clamping down mold halves. An injection nozzle and a screw housed in a heated barrel assembly make up the injection mechanism

Chapter 2

INTRODUCTION TO MICROPLASTICS

Contents

Introduction 28
2.1. Microplastics 28
2.2. Effect of Microplastics 31
2.3. Surveys 33
2.4. Sampling Aspects 34
2.5. Soils 35
2.6. Samples and Their Processing 36
2.7. Techniques for Analyses 39
2.8. Processing 41
2.9. Reports From Research 43
2.10. MP, NP, and HDPE 44
2.11. Analytical Techniques of Microplastics 46

INTRODUCTION

Scientists, decision-makers, and the public at large have grown increasingly concerned about plastic pollution and its effects in recent years. Plastic has grown ubiquitous in all aspects of the environment due to the rapidly increasing business and usage of these materials. As a result, plastic particles have been found in isolated, populated locations or far-off places like Antarctica (Figure 2.1) and the Mariana trenches.

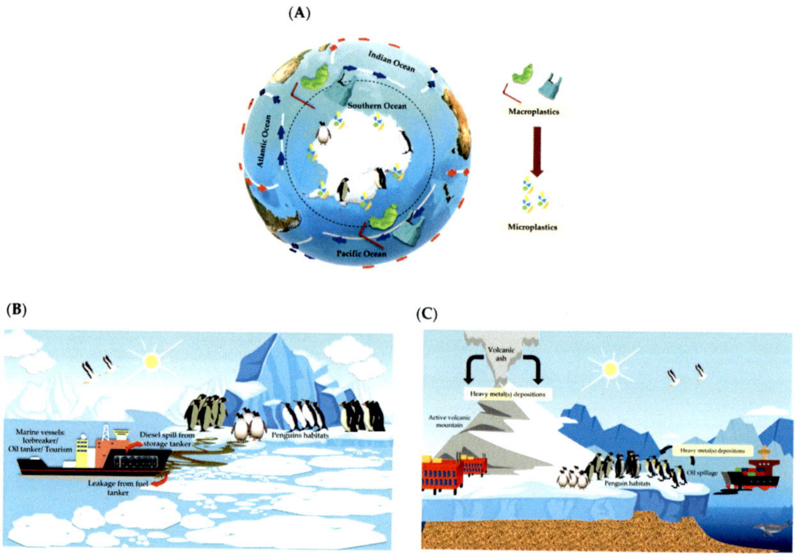

Figure 2.1. *Plastic pollution in Antarctica.*

Source: Puasa NA, Zulkharnain A, Verasoundarapandian G, Wong C-Y, Zahri KNM, Merican F, Shaharuddin NA, Gomez-Fuentes C, Ahmad SA. Effects of Diesel, Heavy Metals and Plastics Pollution on Penguins in Antarctica: A Review. Animals. 2021; 11(9):2505. https://doi.org/10.3390/ani11092505, Creative Commons License.

This is hardly surprising, given that more than 359 million plastics were produced worldwide in 2018; of them, almost 40% were meant for packaging and, consequently, for rapid or almost immediate disposal (Filho et al., 2021). Therefore, plastic pollution is pervasive.

2.1. MICROPLASTICS

Although the effects of this pollution are more noticeable, like when seabirds or whales ingest plastic materials or when seals or sea turtles become trapped, smaller particles, also known as microplastics, may be more common and pose a greater risk to the health of the global environment and, ultimately, to human health. It should be noted that different size-based definitions, ranging from 1mm<=10mm, have also been proposed for microplastics, which are generally classified as plastic particles less than 5 mm.

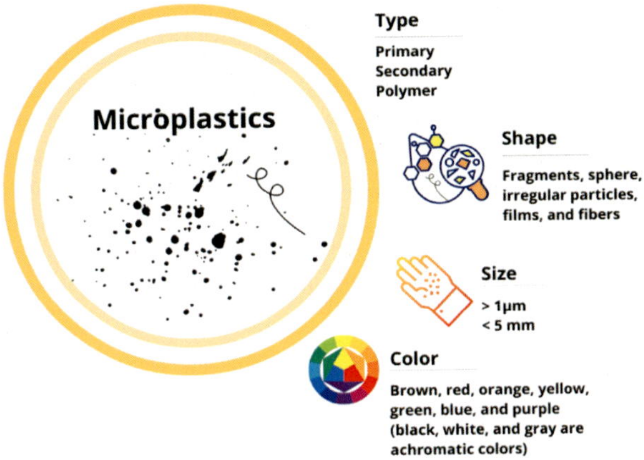

Figure 2.2. *Microplastics.*

Source: Yu, R.-S., & Singh, S. (2023). Microplastic Pollution: Threats and Impacts on Global Marine Ecosystems. Sustainability, 15(17), Article 17. https://doi.org/10.3390/su151713252, Creative Commons License.

Microplastics (MPs) (Figure 2.2) are categorized as primary or secondary MPs based on how they end up in the environment, either directly from their release into the environment or by the fragmentation of bigger particles. All of the risks that are currently being described have been known since the 1970s, when Carpenter and colleagues first reported the presence of microsized plastic particles in the environment and promptly identified the associated hazards and potential consequences.

This is because the scientific community has focused a great deal of attention on identifying, quantifying, and determining the potential effects of these materials once they enter the environment. These included not only the consumption of these materials by a variety of organisms, as is the case with bigger plastic waste, but also the probable adsorption of POPs (permanent organic pollutants) present in water and the microbial colonization of these particles. The use of various additives to improve the properties of plastic materials like malleability, heat and light stability, antioxidant properties, and mechanical resistance has increased the range of applications for these materials, but it has also increased the risks associated with them. One such risk is the leaching of these additives into the environment and biota.

It should come as no surprise that a lot of data regarding microplastics, that is, their frequency, risks, and impacts has been compiled and made public in recent years. However, there are obvious differences and inconsistencies about the existence of these materials in the environment, as well as a noticeable lack of uniformity in the various field studies that have been published. It is evident that the development and implementation of standardized protocols for microplastic sampling, quantification, and characterization along with data treatment and visualization, as well as quality control assurance, are urgently needed to accurately assess the true hazards of microplastics. Only then will it be feasible to reconcile data collected from various research and compare them, allowing for an accurate assessment of the potential impacts of MPs. The choice of sampling location, while taking into account the kinds of samples to be gathered and their preservation, the regional and temporal variation, as well as the available sampling methodologies, is the foundation of any useful and representative study. These materials can change chemically and physically in the environment, and there are currently no defined techniques for sampling them. Moreover, background contamination should constantly be taken into account during the process because of airborne fallout. Therefore, it would make sense to believe that standardized procedures for sampling and sample management actually exist.

In the larger scientific process of finding answers to particular problems, sampling is an important stage. Because of this, the fundamental objective of a sampling method or approach should be its intrinsic reliability, linked to the ensuing stages of analysis and data reporting. Sampling techniques used in Microplastic (MP) research can range from the use of buckets to the intricate and sophisticated usage of oceanographic equipment (Razeghi et al., 2021). In the beginning, sampling for MP was not targeted, and results were frequently derived from samples that had already been collected to analyze sediments and plankton. As a result, the fundamental scientific questions, hypotheses, and sampling designs were not intended to determine the prevalence of these materials but rather served as a complementary set of data. While the scientific community and the general public have long known that microplastics exist in the ocean, there has been a growing interest in the sources, and impacts of these materials once they enter the ecosystem. The number of studies examining the prevalence and consequences of microplastics has increased exponentially as a result of the new status of these particles. Strong regulatory mechanisms aimed at reducing the presence of microplastics in the environment were quickly developed and put into place as a result of the overall findings, which detailed the presence of these materials in otherwise expected pristine environments, such as isolated mountains, the Arctic, the Antarctic, or deep-sea. However, the effectiveness of these mechanisms has been questioned. The ongoing search for regulated, logical, and consistent methods for the collection, analysis, and data reporting of MPs in the environment was nevertheless sparked by such activities.

Different approaches are required to handle different concerns, and this holds true for both the sampling procedure and the analytical techniques applied to a given problem. The sampling plan ought to be in line with and representative of the inquiries that the program's goal is to highlight. Since these materials are naturally found in soils, waterways, and even the air, various matrices require the utilization of suitable and statistically representative samples. The sampling plan that was subsequently developed, along with every other analytical step, should be carried out in a working environment that uses as little plastic as possible or, ideally, is plastic-free. This encompasses, if feasible, all laboratory apparatus since studies have indicated that self-contamination could be responsible for as much as 15% of detected microplastics in ambient samples. It is best to avoid using products made of plastics that are not being evaluated, even though doing so may be judged acceptable. Laminar flow hoods are also advised for handling materials in the lab, especially when handling fibers, which are common as airborne microplastics. In fact, because of this understanding, blanks must be included in order to account for this possible path of contamination. This is frequently supplemented with the use of wet filters housed in borosilicate Petri dishes. These act as contamination controls, reflecting possible airborne microplastic contamination in particular, and are then processed and characterized in a manner that is comparable to the samples that are being studied. These controls must be carried out three times to increase the accuracy of model parameter estimations and enable a statistically meaningful outcome.

2.2. EFFECT OF MICROPLASTICS

Naturally, the matrix from which the microplastics are isolated also significantly influences these effects; for this reason, sample analysis and sampling should always be conducted within the appropriate medium. Even though research on the occurrence and impacts of microplastics has primarily focused on the marine environment, it remains important to understand the fundamentals of how these materials function in the various sections of the marine environment. Microplastics can directly enter the marine ecosystem along the coast or through sea-based activities like fishing.

Microplastics are also present in marine waters due to untreated or inadequately treated discharges from storm drains, residential wastewater, and industrial wastewater. Additionally, rather than entering the environment directly as primary microplastics, microplastics can also form in the environment through the fragmentation of larger plastic waste, known as secondary microplastics. It is essential to trace their origin so that the sample accurately reflects their presence.

Figure 2.3. *Storms transport microplastics.*

Source: Alava JJ, Tirapé A, McMullen K, Uyaguari M, Domínguez GA. Microplastics and Macroplastic Debris as Potential Physical Vectors of SARS-CoV-2: A Hypothetical Overview with Implications for Public Health. Microplastics. 2022; 1(1):156-166. https://doi.org/10.3390/microplastics1010010, Creative Commons License.

Storms and tides can cause microplastics to wash ashore and then wash off again (Figure 2.3) (Zhang, 2017). Plastic particles can be consumed by species of varying trophic levels, which can then carry them through food webs, or they can be swallowed as comparatively dense manure, which sinks to the ocean floor. Biofouling, or the progressive accumulation of organic materials and microorganisms, can also occur. This can change the density of the polymer and impact its sinking behavior. Hetero-aggregation is another phenomenon that can happen and is often dependent on the kind of polymer. Once on the seafloor, these materials may be resuspended due to a process known as bioturbation, which is the change of soils and sediments by plants or animals. The vertical and horizontal dispersion of microplastics in the maritime environment is further complicated by winds, currents, and weather patterns. When choosing a temporal sampling strategy in addition to a sample methodology, all these considerations add to the complexity of the process. In fact, time, money, and spatial constraints frequently limit fieldwork, and seasonal or interannual variations in microplastic sampling are either ignored or minimized, or at most, they are only observed during a given research project, which is typically equal to or shorter than five years.

2.3. SURVEYS

Locations ought to be carefully chosen as well. Sampling surveys from fixed platforms are more limited than specialized research vessels conducting monitoring programs. Samples can also be collected along the seaside; depending on the study subject being addressed, both methods are equally valid. This could be strongly related to other details or interesting aspects, like nearby possible sources of microplastics, like shipping lanes, the existence of industrial facilities, nearby fishing spots, wastewater outlets, tourist destinations, river inflows, or even particular geographic features. Many kinds of equipment are available, depending on the collecting site and the depth of the water to be tested. A surface net tow is the most widely employed technique; it can be carried out with a manta trawl, a neuston net, a catamaran, or bongo nets. Additional techniques might be applied to samples that were obtained in the water column since they frequently contain fewer microplastics. Some authors identified a significant number of microplastic aggregates in the surface microlayer, which has also been specifically sampled using metal sieve meshes and produced microplastic concentrations significantly higher than those found when using other sampling techniques.

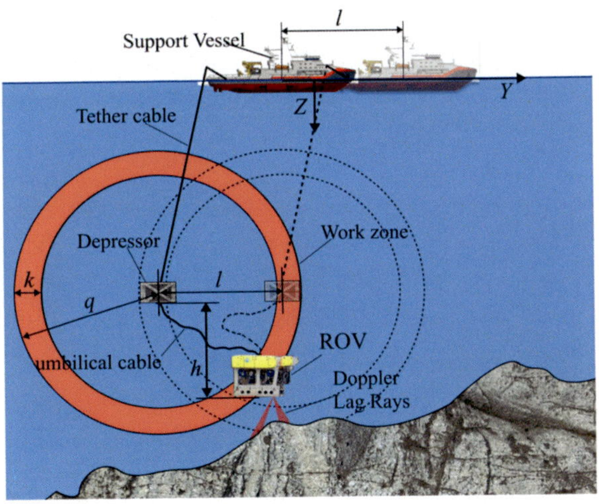

Figure 2.4. *Remotely operated vehicles.*

Source: Konoplin A, Konoplin N, Yurmanov A. Development and Field Testing of a Smart Support System for ROV Operators. Journal of Marine Science and Engineering. 2022; 10(10):1439. https://doi.org/10.3390/jmse10101439, Creative Commons License.

Remotely Operated Vehicles (ROVs) (Figure 2.4) are less common but still an option. By using these vehicles, a more automated sampling technique at various depths may be possible. Additional sensors, such as those measuring pH or suspended solids, may also be used, which could enable a more thorough characterization of the surrounding media and, in turn, a better understanding of the interactions that may have an impact on the distribution of microplastics throughout the water column. These diverse methods make it more difficult to analyze and compare experimentally gathered data thoroughly, especially because each organization, team, or institution has a different method for implementing data reporting. Therefore, using more than one approach may be possible and perhaps advised for the sampling of microplastics in marine waters. Because they can sample large volumes of surface water, manta trawls or nets, for instance, can be used for large-scale surface water sampling. However, because of mesh-size limitations, bulk water or pump-based sampling can be done concurrently, complementing the trawling process and helping to provide a more detailed characterization of the prevalence of smaller microplastics. In fact, some studies showed that when using in situ pump filtration techniques in place of manta trawl, the pump produced better volume accuracy than the trawl, even though the latter sampled higher volumes and concentrations, probably because it sampled particles on the sea surface microlayer more effectively. Therefore, standardizing aperture size, sample period, and trawling speed is necessary to acquire comparable data. Moreover, conducting numerous samplings with changing temporal and geographical characteristics is recommended.

2.4. SAMPLING ASPECTS

Freshwater and estuarine water sampling are commonly regarded as comparable, if not identical, to marine water sampling; they are collectively referred to as aquatic sampling. Nevertheless, rivers have been identified as the main source of microplastics entering the marine environment (van Wijnen et al., 2019). The majority of microplastics must travel through estuaries, which are extremely complex environments influenced by both freshwater and saltwater, before they can enter this compartment. The complexity of estuarine litter sampling and monitoring is further compounded by the bidirectional flow, which alternately flows into and out of the estuary in accordance with tidal cycles. Freshwater systems also include lakes, lagoons, and streams that could be significantly influenced by large-scale microplastic inputs from a variety of human activities, such as urban impacts, waste disposal, agriculture, and fisheries. Each unique geomorphology adds to these complexities by potentially influencing the amount of microplastic deposition along a river or lake. These variables include the transects width and depth, sinuosity, bottom gradient, degree of anastomosis, and the kind and presence of vegetation along the riverbanks. Finally, it is important to take into account the possibility of uneven microplastic distribution in rivers that are restricted in their flow by dams and barrages. However, with the appropriate adjustments, all of these environmental compartments can be sampled using methods akin to those developed for marine waters. Due to the high cost and logistical demands of sampling in the polar regions, data distribution is only unreliable and sampling can be inconsistent. Although samples have been taken in other places, like the Tibetan Plateau and the Alpine Glaciers, there are still few reports on the buildup and discharge of microplastic in glaciers. The time component of the research on microplastics can be enhanced by sampling glaciers, which can reveal atmospheric deposition in polar or high-altitude environments as well as how ice may act as a temporary sink for microplastics and promote their transit.

2.5. SOILS

Soils have only recently been the focus of research to determine the occurrence and possible impacts of microplastics, although for a long time the majority of microplastic studies concentrated on the marine environment and then progressively moved toward freshwater systems (Figure 2.5). Soils are a highly complex matrix that include organic matter, from 0.02% in desert soils to over 100% in marsh soils (Hohl and Varma, 2010).

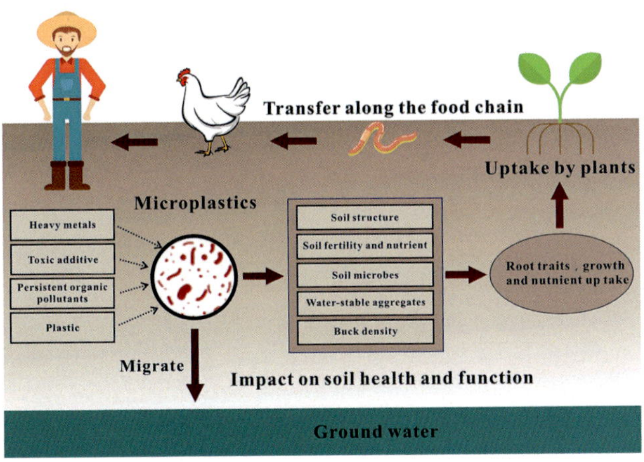

Figure 2.5. *Soils.*

Source: Yang, L., Zhang, Y., Kang, S., Wang, Z., & Wu, C. (2021). Microplastics in soil: A review on methods, occurrence, sources, and potential risk. Science of The Total Environment, 780, 146546. https://doi.org/10.1016/j.scitotenv.2021.146546, Creative Commons License.

Microplastics have the potential to significantly affect soil biophysical characteristics, such as bulk density and water-holding capacity, as well as microbial activity and the development of water-stable aggregates, because this organic matter may interact with other soil components, such as minerals. Sampling procedures need to include both the expected local distribution pattern and the properties of the soil or sediment being analyzed. The spatial distribution of sample sites within an area can be ascertained in a variety of methods, either randomly or following a prearranged procedure, such as a zigzag pattern. If prior data is available, the sampling strategy should take into account the facts that are currently known about contamination. The selection of sampling techniques is mostly determined by their economic proportionality. While it is feasible that sampling procedures have no bearing on the results that follow, there are currently insufficiently detailed studies to support this theory, and it is also plausible that various sample strategies produce varied outcomes.

Another significant factor is the sample volume, which depends on the largest grain size that has been detected. It is not always the case that soils are perceived as the three-dimensional formations that they are. Several papers describe sample techniques that only use surface sampling, or they provide an indicative range for subsurface sampling, such as 2–3 or 3–6 cm. This is extremely important because numerous studies detail the vertical movement of microplastics in soils caused by weather, animal activity, and human activity. Consequently, the amount of microplastic particles per dry weight may be overestimated or underestimated as a result of an inadequate or poorly stated sampling technique. Specific rules and/or standards do exist for other types of solid samples, such as secondary fertilizers like compost, sewage sludge, and biowaste, regarding the quantity of samples to be collected and their corresponding volumes. These can be a good place to start when creating standardized procedures, but they do not account for the sampling of materials like microplastics, so modifications to these suggested methodologies might be necessary. Currently, agricultural soils or sediments are the focus of the majority of investigations on solid matrices. Therefore, additional research is required to determine the frequency of microplastics in various soil types, such as urban and industrial soils, among others. As microplastics have already been shown to be ubiquitous, soils in remote areas should be routinely sampled in the same way as all other sites, even the ones that are sampled the most frequently. This is because a multivariate analysis is required to identify temporal or spatial patterns in contamination.

2.6. SAMPLES AND THEIR PROCESSING

Although there has been evidence of microplastic fallout and atmospheric transport for some time, there is still a lack of research on the topic (Akdogan and Guven, 2019). However, it is evident that the atmosphere plays a significant role in contaminating otherwise isolated and uncontaminated areas. While rain samplers and, in certain situations, plain huge glass containers were employed as non-specific collection equipment at first, more recent developments have led to the development of specialized goods, such as the fallout collector developed by the Norwegian Institute for Air Research (NILU). Constructed from stainless steel, it can be employed in microplastic investigation and ought to be regarded as a conventional passive sample technique for air deposition. These are inexpensive, simple-to-use instruments that don't need energy or complicated infrastructures, making them ideal for usage as uniform sample procedures. Active sampling techniques have also been employed and are thought to be quite effective, especially when monitoring known air volumes over predetermined lengths of time. Because these sampling techniques need established procedures, collected data can be associated with location, weather, and seasonality. Although it needs energy inputs, specific infrastructure, and equipment, it is very reproducible and provides accurate air sample volume determination.

It is important to remember that active and passive sampling techniques should be used in tandem rather than as alternatives. While active sampling techniques enable the identification of microplastics in air masses which may contain a portion of particles that do not deposit, passive techniques offer an overview of the rate of microplastic deposition within particular settings (location and time). Combining sampling approaches with detailed sampling protocols that include information about meteorological conditions, sampling locations, elevation, equipment used, and sampling duration will produce comparable data representative of microplastic deposition as well as the role of scavenging processes, like rainfall, to be quantified. Several studies describe the existence of microplastics in a wide range of taxa, including plankton. Therefore, in freshwater habitats, gill nets are frequently an alternative for collecting organisms, and traps may be used, depending on the research issue at the trophic level under analysis. Trawls at varying water depths are an option for fish, and neuston or bongo nets may be utilized for planktonic invertebrates. It is possible to manually gather other animals like earthworms, bivalves, and crabs. Additionally, organisms can be purchased commercially, for instance, in fish markets. However, there is limited control over possible paths of contamination that may occur between the time a specimen is captured and the time it is analyzed. Capturing larger organisms like dolphins, whales, or sea turtles, is not practical nor ethically acceptable. Nevertheless, there is a way around this restriction given that replicates are required to provide statistically meaningful data: analyzing animal carcasses or gathering and examining animal feces. Organisms should be desiccated, frozen, or preserved in fixatives, such as ethanol or formalin, as soon as possible after collection (Schiaparelli et al.,

2016). This is because defecation can happen anywhere from 30–150 hours after collection, and the stress of being caught can cause regurgitation, which can interfere with the analysis of ingested microplastics. Additionally, weight and size, as well as sex and sexual maturation, should all be documented whenever feasible. It should be mentioned that no organisms have yet to be suggested or cited as a model organism for determining whether or not microplastics are present in various environmental matrices. However, different organisms at different trophic levels, especially those of commercial and economic value, should be identified as organisms of special concern due to the possibility of trophic transfer and, consequently, the chance of human consumption of these materials. It should be noted that while microplastics are more likely to spread throughout the digestive tracts of all organisms, whether aquatic or terrestrial and that they are often discarded rather than consumed, especially in fish, this is not the case for other organisms like bivalves. Furthermore, it is noteworthy that microplastics do not only accumulate in the digestive tracts (Figure 2.6) but also various other tissues and organs such as the testis, liver, and blood.

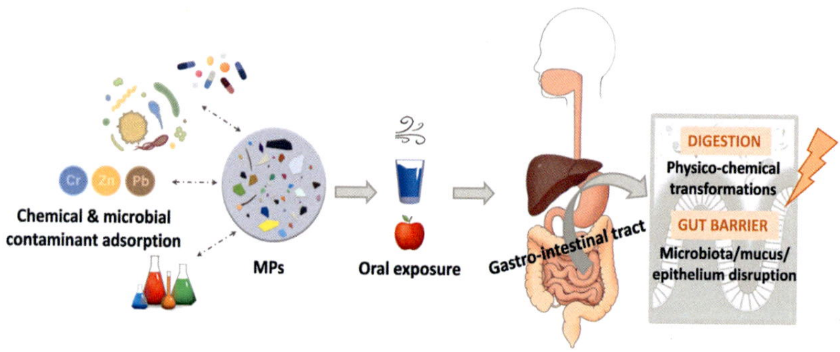

Figure 2.6. *Microplastics in the digestive tract.*

Source: Fournier, E., Etienne-Mesmin, L., Grootaert, C., Jelsbak, L., Syberg, K., Blanquet-Diot, S., & Mercier-Bonin, M. (2021). Microplastics in the human digestive environment: A focus on the potential and challenges facing in vitro gut model development. Journal of Hazardous Materials, 415, 125632. https://doi.org/10.1016/j.jhazmat.2021.125632, Creative Commons License.

Moreover, it has been reported that the brains of certain organisms contain small microplastics, suggesting that these materials may be able to cross the blood-brain barrier (Kopatz et al., 2023). It is necessary to remove microplastics from all other matrix components, including any organic and inorganic materials present, after obtaining a soil, biological, or water sample. However, the specific selection and order of the treatment process that the sample needs to go through is dependent upon the environmental matrix that is being studied as well as the methods of detection and/or identification that are selected.

There are no universally applicable recommendations because the matrix's composition changes depending on the environmental compartment from which it was sampled. However, in order to assess any degradation phenomena through the analysis of various polymer types, morphological features such as shape and size, and weight, it is highly recommended that the separation and extraction steps be accompanied by the use of appropriate reference materials and mixtures. Recovery rates should also be determined to quantify any losses resulting from the separation and extraction methodologies chosen.

Samples are first put through a volume reduction or separation process, like filtering, employing nets or sieves. Bulk samples are occasionally brought into the lab, especially solid materials like soils, biota, and sediments. Similar to sampling, there are no established methods for these processes, despite growing attempts by independent research teams and regulatory and standardization bodies to achieve this goal. However, many methods for separating microplastics have been reported by several authors and are widely acknowledged as standard practices. Not all studies provide concurrent evaluations of the extraction and recovery efficiencies of the selected method, and most methods for cross-calibration, intra and interlaboratory validation, and health and safety have not yet been developed and put into place. Primary assessments are typically carried out through visual inspection, regardless of the kind of sample , that is, soils, sediments, water, or biological. Bulk samples frequently contain detritus, such as pebbles, shells, wood, and/or plants. This means that a visual examination phase is required when the samples are brought into the laboratory. Following optical identification, the microplastics can be sorted according to size, shape, color, and quantity using binocular microscopy, high magnification fluorescence microscopy, or basic visual observation. Though more complex samples may require more rigorous and in-depth analyses, the visual identification and quantification of microplastics in environmental samples can be contentious because it is greatly influenced by the operator, quality, and magnification of the used equipment as well as the sample matrix. Meanwhile, visual sorting takes a lot of time and can only handle a certain size of particle since it can only identify and quantify smaller particles that are manageable in terms of size.

2.7. TECHNIQUES FOR ANALYSES

Regardless of the methodology selected, samples should always be handled and processed in the laboratory to prevent any needless contact between the samples and the surrounding air. This is because the more steps in the methodological protocol chosen, the higher the likelihood of contamination, especially by airborne fibers. Additionally, a realistic evaluation of the expenses associated with chemicals and their associated toxicity should be conducted, particularly when processing high volumes or numbers of samples. The process of plastics rising to the surface of a denser solution, which is based only on density differences, is referred to as flotation in literature when discussing density separation. As a result, the hydrophobicity of the materials also affects froth flotation in addition to density. In a nutshell, variations in the generated air bubbles' capacity to stick to the surface of the particles of interest in a way that depends on their hydrophobicity cause the particles to be physically separated from the liquid phase. As more hydrophilic components remain in the solution, the hydrophobic particles connected to the air bubbles are carried to the surface, generating a froth that is readily removed. In the case of a plastic mixture, separation can be accomplished by adding a wetting agent that preferentially adsorbs to one of the materials, making it more hydrophilic. Nevertheless, flotation necessitates the assessment of a number of parameters, such as the particles' surface free energy, the liquid's surface tension in the flotation bath, and the critical surface tension. Though good recovery rates have been observed, the approach is not commonly employed for, say, sediments since it has been found that the use of wetting agents or additives prevalent in plastics severely affects the procedure.

Elutriation is a technique wherein a liquid, like water, is injected at the base of a column to fluidize the sample and facilitate the separation of microplastics that are already present (Kedzierski et al., 2016). Elutriation may make it possible to separate the materials according to their density, size, and shape, depending on how complicated the system is. After that, high density solutions are used to gather the particles into a mesh and segregate them. Low-cost elutriation makes it possible to separate microplastics from bigger samples very effectively, increasing their environmental representativeness. The Munich Plastic Sediment Separator (MPSS) operates on the elutriation principle, wherein a concentrated salt solution is injected at the bottom of the column, and the particles that rise are either directly vacuum-filtered or collected in the supernatant. Nevertheless, despite the greater extraction efficiency for reference particles (>97%), it has subsequently been shown that the MPSS may only achieve an efficiency of 13–39%, in addition to the lengthy processing time. Perhaps the most popular technique for separating microplastics is filtration, which is also usually included as a necessary step after other separation procedures like density separation. Although the method is simple, processing samples with particularly high amounts of particulate matter may cause complications. It is a highly effective physical separation process in which the size of the retained solids is defined by the pore size and, for aqueous samples, the technique is commonly based on the use of a funnel, filter paper, and a vacuum system. This causes the filter to clog quickly, which lengthens the time needed to process samples and decreases process efficacy.

2.8. PROCESSING

Of course, adding more stages to a sample increases the likelihood of losing some of the particles in it, which could lead to an underrepresentation of the prevalence of microplastics in that sample. The filter paper's pore size varies a lot; reports have shown pore dimensions ranging from 0.3–200 μm, however, most often fall between 1 and 2 μm. The most crucial factor, albeit it is less often taken into account, is the filter's composition. Glass fiber and cellulose ester filters are two of the most widely used materials; however, low fluorescence materials, including aluminum-coated polycarbonate membrane filters, are advised if aiming for future spectroscopic investigation, such as Raman spectroscopy. Additionally, before using filters, they should be examined under a microscope for contamination. In order to make sure that any possible microplastics are not taken into account in the analysis that follows, a prior procedure like calcination may be used. Some less common approaches have been documented in the literature, while the previously described methods are the most commonly employed in the analysis, identification, and quantification of microplastics. Among these are electrostatic separators, such as the bell separator made by Korona-Walzen-Scheider (KWS) (da Costa et al., 2022). Despite being quick, this approach produces extremely varied recovery rates, which can range from 45% to 100%. The main factor influencing the observed irregularity in recovery is the size of the separated particles. Therefore, internal standards must be used while utilizing the KWS separator, especially when evaluating tiny particles. Imaging flow cytometry has become a viable alternative to produce quantitative high-throughput imaging data by fusing flow cytometry with digital microscopy. This technique may find use in the field of microplastic research. But it is still an unvalidated technology whose wider application is hindered by the high cost of core facilities and equipment, the difficulty of analyzing and interpreting image-based data, and the requirement for highly skilled workers. It is also an extremely time-consuming process that prevents the measured polymers from being characterized.

Researchers proposed an innovative method based on particle-impact electrochemistry. These scientists accurately determined the concentrations and particle size distributions of spherical polyethylene (PE) microparticles by analyzing the impact frequency of microplastics in solution with an electrode, driven by Brownian motion. Although this approach is praised for its high accuracy, it is currently limited to pristine microplastics, as it assumes that the polyethylene microparticles are electrochemically inactive within the applied potential range. The 1970s saw the first reports of tiny plastic particles in seawater and on beaches—an early indication of a worsening environmental issue. Alarmingly, an estimated 6.3 out of the 8.3 billion metric tonnes of plastic debris produced since the 1950s has entered waterways and oceans, contaminating water supplies. The primary concern with plastic litter is its rapid transformation into the global challenge of microplastic and nanoplastic contamination. Plastic particles smaller than 5 mm, known as microplastics, can be morphologically classified as films, foams, fibers, spheres, or pellets. The term "plastic" encompasses a wide variety of synthetic, water-insoluble polymers, including, but not limited to, rubber, polyacrylonitrile, polyamide (nylon), polyethylene (PE), polyethylene terephthalate (PET, polyester in fiber form), polypropylene (PP), polystyrene (PS), and polyvinyl chloride (PVC). Primary plastic products photodegrade when exposed to ultraviolet light, breaking down into smaller fragments called secondary plastics. Additionally, microplastics can originate from primary sources, such as cosmetic microbeads and microfibers released by synthetic textiles. Microplastics and nanoplastics, with their higher surface-to-volume ratios, tend to settle more rapidly on the ocean floor (Yu et al., 2019). It is well-known that planktonic organisms ingest microplastics, introducing them into the marine food chain when consumed by marine life. Furthermore, the bioaccumulation (Figure 2.7) of toxic substances and contaminants adsorbed on the surface of microplastics has harmful effects on aquatic organisms and marine ecosystems, potentially leading to species extinction.

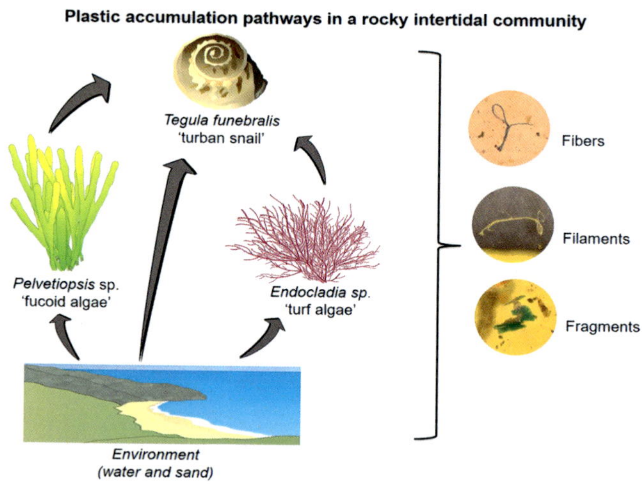

Figure 2.7. *Bioaccumulation of microplastics.*

Source: Saley, A. M., Smart, A. C., Bezerra, M. F., Burnham, T. L. U., Capece, L. R., Lima, L. F. O., Carsh, A. C., Williams, S. L., & Morgan, S. G. (2019). Microplastic accumulation and biomagnification in a coastal marine reserve situated in a sparsely populated area. Marine Pollution Bulletin, 146, 54–59. https://doi.org/10.1016/j.marpolbul.2019.05.065, Creative Commons License.

The full scope and long-term effects of the chemical, physical, and biological impacts of microplastics remain unknown, while environmental littering continues unabated. Given the prevalence of microplastics, it has been established that humans are also ingesting and inhaling them.

2.9. REPORTS FROM RESEARCH

According to several studies, microplastics have been discovered in food and water supplies, the air, and even the rain. Global reports on microplastic contamination of natural aquatic habitats have sparked a broad and focused discussion about microplastic sampling techniques and the need for standardization. There is a critical lack of harmonization between methodologies and units in the accurate measurement of the quantitative distribution and composition, including the kind of polymer, its nature and origin, as well as the sources and pathways of microplastics. The outcomes of the techniques used in microplastic analysis are therefore significantly influenced by the sampling, pre-treatment, separation, and post-treatment procedures.

Since visual inspection is a crucial initial step in conducting an accurate evaluation of such materials, microscopy remains the method for the identification and characterization of microplastics. Microscopy will always be the first method used for evaluating microplastics, regardless of any controlled standard process for the sampling and sorting of microplastics before their examination. Visual inspection is an easy way to identify macro- in terms of the size range, however, imaging techniques are needed to recognize microplastics.

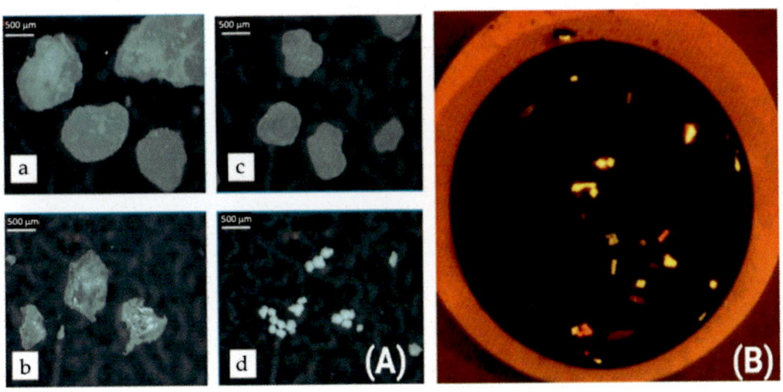

Figure 2.8. *Optical microscope images of microplastics.*

Source: Sarkar S, Diab H, Thompson J. Microplastic Pollution: Chemical Characterization and Impact on Wildlife. International Journal of Environmental Research and Public Health. 2023; 20(3):1745. https://doi.org/10.3390/ijerph20031745, Creative Commons License.

Consequently, microscopy is an essential analytical method for identifying and measuring microplastics in aquatic biota and water samples since it gives vital details about the shape, size, and chemical makeup of these micromaterials.

An apparatus with one or more lenses that magnifies an object of interest placed in the focal plane is known as an Optical Microscope (OM) (Figure 2.8). OM has two primary imaging operation modes that vary based on the light path's beam. When observing thick and opaque materials, the light beam is reflected off the sample surface, resulting in contrast due to variations in reflectivity and topography. Global plastic output has increased significantly during the last ten years, reaching 359 million metric tonnes in 2018 (Mai et al., 2020). The scientific community and the public media have grown more concerned about the long-term accumulation of plastic garbage due to the extraordinarily slow rate at which it completes mineralization under environmental conditions.

2.10. MP, NP, AND HDPE

Microplastics (MP) are typically defined as plastic waste particles on the micrometer scale, with a maximum size limit of 1 or 5 mm. Larger operational size limits, such as 10, 100, or 300 μm, are sometimes utilized due to sampling or analytical limitations, but the minimum size of MP is typically stated as 1 μm. Although a consensus or standard definition of nanoplastics (NP) has not yet been fully established, a proportion smaller than MP, or less than 1000 nm, is most frequently advocated. It was only recently suggested that NP existed and that they may be harmful to the ecosystem while having previously gone unnoticed. Research is still being done to determine their frequency and impact. MP can be divided into primary microplastics and secondary microplastics, depending on their origins. Primary microplastics are primarily found in paint, composite materials, medications, and personal care items. They are intentionally created and produced as tiny particles. Secondary microplastics are produced when larger plastic waste breaks down and weathers as a result of mechanical wear and tear, UV light, and microbiological deterioration in the environment. The microplastics are thought to be secondary MP, which come from household objects, packaging, single-use goods, and other discarded plastic waste. Consequently, there will be more secondary MP in the environment as a result of the increased intake of plastic debris from these sources. Secondary microplastics also include fibers that are discharged during the washing of synthetic clothing and that get into aquatic environments through sewage treatment facilities. Because MPs are so common in both aquatic and terrestrial environments, even if their effects on public health and aquatic ecosystems are still not entirely known, it is anticipated that MP will be a significant risk. The presence of MPs in the open ocean was first documented in the 1970s, and in more recent times, researchers have focused more on the substances prevalence and detrimental impacts on freshwater and marine biota. Their small size allows it to be ingested by a variety of marine species with varying trophic levels and feeding preferences, which can result in both direct physical harm and possible toxic effects. Physical damage specifically refers to MP's inherent mechanical ability to cause injury to digestive tracts. There is a chance that the nutritional and lipid condition of the organisms will be compromised at high plastic occurrence and ingestion rates relative to organism clearing rates. Furthermore, several theories have been put forth concerning the effects of co-contaminants with plastics. These include the possibility that plastic additives or monomers could seep into the organisms or that plastic particles will enhance the concentration of hazardous and persistent organic pollutants (POPs) in the surrounding environment, serving as a vector for chemical uptake (Rios et al., 2007). Even though these consequences have been qualitatively proven, there is ongoing debate over how much of an environmental concern this should be. Various seafoods have been linked to these risks (Figure 2.9).

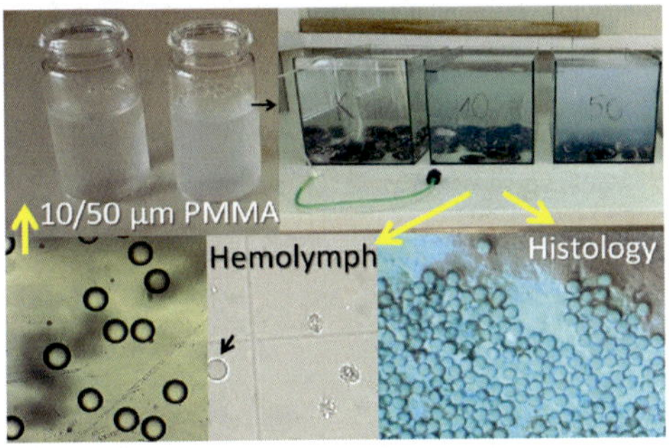

Figure 2.9. *Microplastics in the mussel Mytilus galloprovincialis.*

Source: Pavičić-Hamer D, Kovačić I, Sović T, Marelja M, Lyons DM. Exposure to Polymethylmethacrylate Microplastics Induces a Particle Size-Dependent Immune Response in Mediterranean Mussel Mytilus galloprovincialis. Fishes. 2022; 7(6):307. https://doi.org/10.3390/fishes7060307, Creative Commons License.

For instance, High-Density Polyethylene (HDPE) particles with sizes ranging from 0 to 80 µm were consumed and absorbed into the cells and tissue of the mussel, *Mytilus edulis* L. After three hours of exposure, they discovered that MPs were absorbed by the stomach and moved into the digestive gland, where they gathered in the lysosomal system. Plastic pollution is gaining a lot of public attention and has sparked calls for policy action because of the possible damage to human health. The United States has prohibited the use of microbeads since 2015, while the European Chemicals Agency has suggested imposing extensive restrictions on the purposeful use of MP in goods sold in the EU in 2019. Following the publication of preliminary findings in 2004 regarding the presence of MP in sediments and seawater, the scientific community worldwide focused more on MP pollution. Since then, scientists have documented evidence of microplastic contamination in seemingly all places, including food supplies, drinks, and lakes and rivers. Despite the recent sharp rise in the number of publications on microplastics, there is still no standardized method for identifying or quantifying MP, and the methodologies employed by various research groups vary greatly, making the data from studies completely incomparable. The majority of sampling, lab prep, and identification/quantification techniques are currently being developed and tested.

2.11. ANALYTICAL TECHNIQUES OF MICROPLASTICS

A variety of analytical techniques are suggested for MP identification, including liquid chromatography, pyrolysis, gas chromatography-mass spectrometry (GC-MS), densitometry, vibrational spectroscopy, and differential scanning calorimetry (DSC).

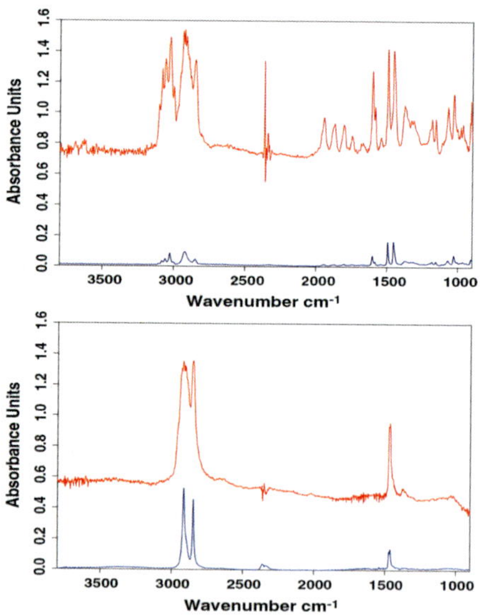

Figure 2.10. *Vibrational spectra of MP.*

Source: Löder, M. G. J., & Gerdts, G. (2015). Methodology Used for the Detection and Identification of Microplastics—A Critical Appraisal. In M. Bergmann, L. Gutow, & M. Klages (Eds.), Marine Anthropogenic Litter (pp. 201–227). Springer International Publishing. https://doi.org/10.1007/978-3-319-16510-3_8, Creative Commons License.

Infrared and Raman vibrational spectroscopies are the most widely utilized among these methods for chemical identification. The vibrational spectra of MP particles (Figure 2.10) are unique for each of the main polymer types and are used by IR and Raman spectroscopies to identify them. Chemical imaging is made possible by the integration of vibration spectroscopic techniques into microscopic setups. Chemical imaging is superior to other analytical techniques because it combines the benefits of digital imaging with the characteristics of spectroscopic measurements to allow simultaneous determination of MP particle counts, sizes, and polymer identity. As complimentary vibrational spectroscopic techniques, infrared and Raman spectroscopies allow for the observation of weak IR intensities in many instances and significant Raman intensities in others. Raman provides greater size resolution for MP analysis, but it is typically a confocal method that requires precise focus in three dimensions, lengthens measurement times, and necessitates the optimization of additional parameters. In theory, IR spectroscopy, the most popular approach for characterizing MPs, provides a quick, effective, and non-destructive way to identify and discriminate the majority of plastic polymers. Nonetheless, a variety of factors must be considered in order to produce an effective, reliable, and clear MP identification.

When light photons collide with matter, several events can occur. Initially, light may be reflected or pass through the substance unaffected. Photons may be absorbed while stimulating molecules, which could result in fluorescence or be employed in IR spectroscopy. There is a chance that the light is scattered; it might be inelastic with a change in photon energy or elastic without one. The Raman effect got its name from the inelastic scattering of photons that C. V. Raman and K. S. Krishnan originally noticed in 1928 (Prochazka, 2016). Chemical compounds can exist in multiple excited electronic states in addition to their ground state, which is their electronic state. A molecule can exist in several vibrational states inside the electronic states. Energy can be absorbed or emitted to alter an energetic state. A molecule is placed in a labile state when a photon is scattered upon it and its energy is absorbed. The molecule returns to its initial vibrational state after emitting a photon with the same energy as the one that was absorbed. Rayleigh scattering, the most likely event, is what it is known as. A lower energy photon is released, causing the molecule to return to a higher energy vibrational state. This process is known as Stokes scattering. The molecule returns to a lower energy vibrational state and emits a photon with an increased energy. We refer to this as anti-Stokes scattering. Stokes, anti-Stokes, and Raman scattering are extremely uncommon. Because anti-Stokes scattering is considerably less common, Raman bands are weaker as a result. Anti-Stokes scattering requires excited vibrational states, which are only found in a relatively small number of molecules at normal temperature.

The molecule, or more specifically, the chemical bonds and vibrational states within, determines the energy differential between the absorbed and released photons. Thus, one can infer the chemical bonds that exist in a molecule based on the amount of energy acquired or lost during scattering. Raman spectroscopy uses this phenomenon to identify and describe materials.

Further, a shift in the polarizability of a chemical bond indicates that an external electric field can more easily create an electronic dipole. As a result, the energy of the dispersed light is significantly more than the chemical bond's vibrational energy. On the other hand, chemical bond vibrations that alter a bond's polarization, or its current dipole moment, are infrared active. In this instance, only infrared light with an energy comparable to the chemical bond's vibrational energy is absorbed. Using infrared spectroscopy, this phenomenon is utilized to detect and characterize chemical substances. While chemical linkages involving the carbon backbone of organic compounds are more Raman active, those involving functional groups are more IR active. As a result, the data obtained using the two methods is considered complementary. Fluorescence is a phenomenon that can occur after light is absorbed. By radiation less emission and photon emission, a molecule can return to its ground state of electronic energy if the energy of an absorbed photon is sufficient to raise it to an excited electronic state. The Raman effect can be utilized in conjunction with Raman spectroscopy to identify and characterize chemical substances and materials. The idea is to light up the sample, scatter the light, gather it, direct it toward a detector, and then convert it into a recognizable spectrum that can be deciphered. Typically, monochromatic lasers with wavelengths ranging from visible to near infrared are used as the light source. The intensity of Raman scattered light is directly proportional to the laser's power and fourth power of frequency. Consequently, it is indirectly proportional to the laser's wavelength raised to the fourth power. Therefore, a stronger Raman effect is usually produced by lower wavelength lasers. Lower laser wavelengths, however, are associated with larger photon energies, which may be sufficient to cause fluorescence.

Since weak Raman bands are either barely visible, nonexistent, or the detector signal is saturated by the fluorescence light, fluorescence itself is one of the main sources of interference in Raman spectroscopy. However, several popular detector types like charge coupled devices, or CCDs are less sensitive in the near infrared spectrum, which poses a challenge for longer wavelength lasers. Light emitted (fluorescence), scattered (Rayleigh and Raman), and reflected from the sample are all collected. Usually, a filter is used to exclude the strong reflected and Rayleigh light that has a wavelength comparable to the stimulating light. For example, slits and holes are utilized to focus the remaining light before it is directed toward the detector. One can examine the scattered light that is Stokes or anti-Stokes. Stokes light is typically examined since it is more intense. The energy lost as a result of both fluorescence and Stokes scattering is equivalent to the difference in the molecular vibrational states. Since the light emitter's wavelength is longer than the illuminator, they are in the same spectral range. This effect cannot occur when examining the anti-Stokes light since the light that is released has a shorter wavelength than the illuminator. After that, the detector creates a Raman spectrum from the collected light. The intensity of the light is plotted against the wavenumber shift with respect to the radiation source. The wavenumber is the reciprocal of the wavelength and is commonly given as cm^{-1}. The shift in wavenumber, which is connected to the energy between the different vibrational states of chemical bonds or molecules, produces characteristic bands in the spectrum. By examining the position and intensity of these bands or by comparing the unique spectrum (fingerprint) with known spectra, it is possible to identify the sample material. Raman microscopy or RM can be created by combining these characteristics of Raman spectroscopy with those of a microscope. The objective is used to capture the dispersed light after the sample has been exposed to radiation. Low-size samples are thus made available. Confocal Raman microscopes predominate.

Chapter 3

MICROPLASTICS AND THEIR CHARACTERIZATION

Contents

Introduction..52
3.1. Plastic Waste 53
3.2. Microplastics-More on Them 54
3.3. Occurrence and Effects...55
3.4. Associated Reactions 56
3.5. Microorganisms-Based Breakdown of Plastics..................................57
3.6. Consumption of Polymers and Effects 59
3.7. Research Findings...62
3.8. Sampling of Microplastics...63
3.9. Collection Details..65
3.10. Problems With Microplastics 66
3.11 Assessing Microplastics in Aquatic Environments............................68
3.12. Identification Reports of Microplastics 71

INTRODUCTION

A variety of semisynthetic or synthetic organic compounds can be molded into a soft shape and then set into a very stiff or somewhat elastic form to create plastics. Plastics is a general term used to describe polymeric material that may incorporate other components to increase performance and/or lower costs (International Union of Pure and Applied Chemistry or IUPAC) (Desidery and Lanotte, 2022). Additional characteristics include low cost, low manufacturing effort, chemical and water resistance, temperature resistance, and light resistance. Due to these qualities, plastics have displaced and replaced a wide variety of materials, including wood, paper, metal, glass, ceramics, stone, and leather. At the moment, plastics are found in an enormous and ever-expanding spectrum of items, from paper clips to spaceships.

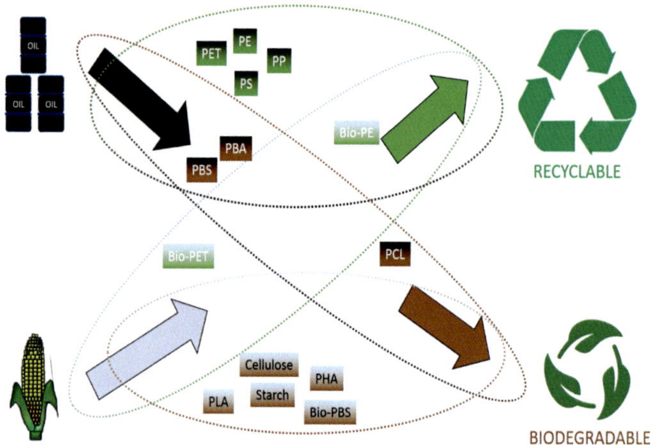

Figure 3.1. *Biodegradable plastics.*

Source: Narancic, T., Cerrone, F., Beagan, N., & O'Connor, K. E. (2020). Recent Advances in Bioplastics: Application and Biodegradation. Polymers, 12(4), Article 4. https://doi.org/10.3390/polym12040920, Creative Commons License.

This has taken different forms, such as thermoplastics, naturally occurring and synthetic polymers, and, more recently, biodegradable plastics (Figure 3.1) in response to growing environmental concerns. It is therefore not unexpected that the comprehensive report on the yearly global output of plastics for 2015 indicated that it exceeded 310 million tonnes given its flexibility. Even though plastics have many advantages, their extensive use, particularly when it comes to disposable goods like packaging, causes them to eventually accumulate in the environment. In fact, it is estimated that plastic debris makes up 10% of all municipal waste produced worldwide (Lebreton and Andrady, 2019).

3.1. PLASTIC WASTE

While some of this plastic debris is recycled, it ends up in landfills, where it can take several hundred years for it to break down. Plastics that end up in the marine environment, which make up around 10% of all plastics generated, are a particular source of concern. The term macroplastics refers to this larger category of plastic waste that has long been the subject of environmental research. This includes certain regions of the ocean where surface current convergence causes them to tend to accumulate, as is the case with the Great Pacific Garbage Patch (Figure 3.2).

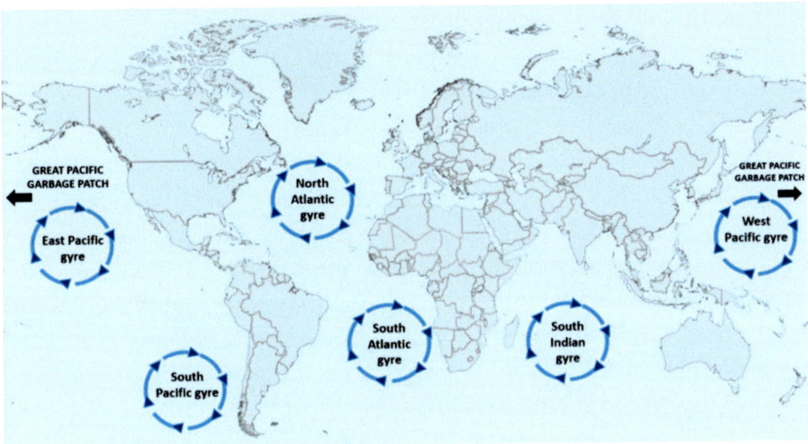

Figure 3.2. *Great Pacific Garbage Patch.*

Source: Leal Filho W, Hunt J, Kovaleva M. Garbage Patches and Their Environmental Implications in a Plastisphere. Journal of Marine Science and Engineering. 2021; 9(11):1289. https://doi.org/10.3390/jmse9111289, Creative Commons License.

The combined forces of wind and waves constantly mix this floating debris, causing it to become extensively distributed over vast surface regions and throughout the uppermost part. These floating debris can also serve as new pathways for invasive species. It is interesting to note that research has also shown that the organisms connected to this plastic waste, referred to as the Plastisphere, can occasionally diverge significantly from the free-floating microbial communities frequently observed in the oceans (Hou et al., 2021). However, there are additional environmental hazards associated with plastic waste that are possibly less well-known. These include threats to shipping, fisheries, and other industries like aquaculture and energy generation. These are by no means the only risks that plastic waste in the ocean poses. Despite the scarcity of research on the socioeconomic effects of this debris, the primary industries impacted have been identified. The multiscalar and temporal mechanisms underlying the fate and behavior of plastics in the environment underscore the significance of this issue, and their effects transcend boundaries.

3.2. MICROPLASTICS-MORE ON THEM

Primary microplastics are produced with a millimetric or sub-millimetric size and are used in toothpaste, lotions, and personal hygiene products as face cleansers (Batool and Qadir, 2022). They are also used, to a lesser degree, in medicine as drug carriers. For instance, it is estimated that microplastics, of which Polyethylene (PE) makes up more than 93%, are present in 6% of the liquid skin-cleaning solutions marketed in the European Union, Switzerland, and Norway. Microplastics found in these products are commonly disposed of after use, and wastewater collection and treatment systems may allow them to enter the environment. These polymeric materials are also sourced from the currently popular 3D printers, which have been demonstrated to produce particles as small as 11.5 nm. The raw ingredients used to make plastic items are another important source of primary microplastics. The environment may also get contaminated by unintentional loss, careless handling, runoff from processing plants, and leftovers from the production of plastic products. When bigger plastic particles break down, secondary microplastics are produced. The physical, biological, and chemical processes that these polymeric particles are subjected to when exposed can result in a decrease in the structural integrity of plastic waste, which can then fragment. But, as is the case with synthetic fibers discharged during laundry, this fragmentation can also happen before the plastics reach the environment. The absence of a common definition and characterization of microplastics, as well as established procedures for sampling, unit standardization, and data expression, further exacerbates these limitations.

Plastic comes from anthropogenic sources primarily found inland. Nevertheless, up to 80% of these materials wind up in the oceans as a result of surface run-off, wind, and discharge (industrial wastewaters). While current mechanical wastewater treatment processes have demonstrated a high degree of efficiency in eliminating microplastics, the effectiveness of these processes varies substantially depending on the materials present and their load, in addition to the types of treatments employed in wastewater treatment plants. Because most plastic materials have low densities, it is believed that a sizable fraction of them wind up in the oceans. Nevertheless, this could not always be the case because these materials could settle in riverbeds due to processes such heteroaggregation with other debris and suspended solids, which has recently been supported by research describing similar occurrences.

3.3. OCCURRENCE AND EFFECTS

Plastic waste is carried across great distances by winds and ocean currents, and it can be found in all areas of the oceans, including the deep ocean, distant islands, and the North and South Poles. Plastic waste can therefore enter marine ecosystems anywhere in the world. According to circulation models, the five subtropical gyres are where this detritus accumulates. These massive vortices function as conveyor belts, gathering the discharged floating plastic waste and accumulating it in central convergence zones. They are maintained by Ekman currents. Direct measurements of microplastics in the surface have confirmed these predictions, despite the extreme complexity of these distribution dynamics.

Nevertheless, it has been discovered that the prevalence of microplastics is significantly higher in coastal areas compared to offshore locations. This finding can be linked to human activity, namely in inhabited areas with high levels of industrial and commercial activity. Plastic debris can be found throughout the water column despite the low density of the most widely used plastics, such as PE, Low-Density Polyethylene (LDPE), High-Density Polyethylene (HDPE), and Polypropylene (PP) (Dimassi et al., 2024). Particles composed of materials having a specific gravity >1 have a tendency to sink when thrown straight into an aquatic environment, whereas particles with a specific gravity <1 tend to stay afloat. However, the densities of these materials may change over time as a result of other phenomena, such as microbial colonization, in addition to the previously discussed heteroaggregation (biofouling). In as little as three weeks, the density of these materials has been shown to be dramatically impacted by the creation of biofilms, which eventually results in the presence of algae and/or invertebrates. These occurrences, along with the intrinsic turbulence of the ocean and the material interactions with the biota, cause plastic waste to disperse throughout the water column, creating such a complex system. Microplastics can degrade in the environment as a result of biotic or abiotic processes that might operate simultaneously or sequentially. Nevertheless, these methods also rely on the physical and chemical properties of the polymeric materials, not just the surrounding conditions. Because of their ester linkages, polyurethanes, for instance, biodegrade more quickly. Additionally, adsorption has been reported to favor positively charged plastic beads over negatively charged ones because of the electrostatic attraction between the charged plastic particles and the components of the living systems. This could also have an impact on the likelihood that microbes would adhere to their surface and create biofilms, which could then have an impact on how quickly the biotic degradation processes proceed. The intricacy of these materials' composition and polymeric structure may also have an impact on their abiotic and biological breakdown processes. Polymers like Polyethylene Terephthalate (PET),

Polypropylene (PP), and Polyethylene (PET) that have short repeating units with high symmetry frequently restrict enzyme accessibility, making them less vulnerable to the action of these biomolecules. As a result, it has been noted that various plastic material blends have varying susceptibilities to Ultraviolet (UV)-mediated deterioration.

3.4. ASSOCIATED REACTIONS

Plastics go through mechanical disintegration, pressure variations, water turbulence, freeze-thaw cycles, and animal damage when they are exposed to weathering conditions. However, this mechanical breakdown is not the same as degradation because the materials only undergo morphological changes instead of molecular bond changes. Most people agree that the most effective abiotic degradation pathway in the environment is photodegradation. Both visible (400–700 nm) and high-energy UV (290–400 nm) radiation can be absorbed by plastics (Andrady et al., 2023). This increases the reactivity of their electrons, causing oxidation and cleavage, which are degradation processes primarily mediated by chain breakdown and cross-linking events. Plastics undergo thermal degradation, which results in bond breakage of the primary polymeric chain and modifications to the material's tensile strength, molecular weight, crystallinity, and even color. Nevertheless, the elevated temperatures necessary for these alterations are hardly noticed in environments where microplastics are present. However, given that samples of PP that have had thermal pretreatment are 25 times more vulnerable to biodegradation than those that have not been prepared, thermal treatment of polymers may still be environmentally significant. Another significant cause of the environmental deterioration of plastic is oxidation. This process results from the introduction of oxygen into the polymer matrix and can be either thermally or photoinduced. Consequently, subsequent biotic degradation pathways benefit from the creation of carbonyl (CO) and hydroxyl (OH) functional groups. Even at low concentrations, ozone (O_3) in the environment speeds up the deterioration of plastics because it is a potent oxidant that breaks down covalent bonds, causing chain scissions and/or crosslinking processes that release free radicals.

Hydrolytic breakdown is an additional process that plastics can go through. However, the presence of covalent bonds that are prone to hydrolysis, like those found in ether and ester groups, is a direct need for this activity. This causes the polymeric chain to break and, as a result, the molecular weight distribution, which alters the strength and other qualities of plastics. Abiotic degradation processes are insignificant in the benthic zone because of the low temperatures, limited rates of solar penetration, and minimal oxidation-mediated mechanisms. However, the ongoing agitation and attrition with sediments produce smaller particles all the time, which build up on the seafloor and provide a persistent source of environmental exposure. Additionally, studies indicated that the high pressures now present in these places may also play a role in their degradation; however, further evidence is needed to support these claims.

3.5. MICROORGANISMS-BASED BREAKDOWN OF PLASTICS

The biological breakdown of plastics typically starts outside of the cell when released enzymes hydrolyze the polymer chains (Shah et al., 2008). These mechanisms occur in soils as well as aquatic environments, and they are further strengthened by the polymer chain's synthesis of smaller units. As their molecular weight continuously decreases, cross-linking reactions which are facilitated by the rising concentration of oxygen and water lead to the continuous loss of the polymeric structure and the formation of molecules that are more vulnerable to microbial action. These interactions produce monomers and oligomers that are soluble in water, which start the mineralization process. These substances can pass through the semipermeable outer membrane of microorganisms and are absorbed as a source of carbon or nitrogen via the proper metabolic route. Certain materials can show significant rates of bio-assimilation, even though they are not fully assimilated. PE, for instance, has been shown to be at least 60% assimilated after 180 days in composting conditions.

Because there are fewer microbial communities in oceans, the rate of biodegradation at the benthic level is thought to be quite small. However, in less deep seas, the numerous microbial communities that already exist, including autotrophs, heterotrophs, and symbionts, actively contribute to the biodegradation of plastics. In fact, these communities have been observed at the surface of marine plastic waste. According to a recent attempt to estimate the load of plastics in the oceans using more than 3000 samples gathered from all over the world, there are between 7000 and 35,000 tonnes of plastic at the ocean's surface (Wu et al., 2019). However, the same authors estimated that more than a million tonnes of floating plastic had been dumped into the open ocean since the 1970s, indicating a discrepancy of up to 100 times between measured and expected levels. As a result, this study highlighted both the enormous loss of plastic at the ocean's surface and the notable disparity in the distribution of floating plastic debris. The rapid breakdown of plastic particles from the millimeter to the micrometer and nanometer size, which produces the so-called nanoplastics, and the ingestion by marine organisms with subsequent egestion are more reasonable causes. The need for more precise and accurate methodologies in the sampling and quantification of microplastics in the environment is highlighted by the fact that high surface-to-volume ratio particles, like microplastics, which are thought to be the most common-sized particles in the ocean gyres, can undergo biofouling processes and display reduced buoyancy as a result. These processes are not taken into account in the sampling procedures. While there may be unexplained sink mechanisms as well, it is more likely that a combination of these activities accounts for the observed disparity.

Figure 3.3. *Landfills.*

Source: Dabrowska, D., Rykala, W., & Nourani, V. (2023). Causes, Types and Consequences of Municipal Waste Landfill Fires—Literature Review. Sustainability, 15(7), Article 7. https://doi.org/10.3390/su15075713, Creative Commons License.

The terrestrial environment can also be contaminated by microplastics, and soils are thought to play a significant role as sinks for these particles. For instance, abandoned plastic items discovered in landfills (Figure 3.3) could be prone to abrasion and attrition, go airborne, and then build up in the soil through atmospheric deposition. The accumulation of these materials in agricultural areas through irrigation with water from wastewater treatment plants that contain fibers and other tiny polymeric components is another possibility. As long as plastics are produced worldwide, there will almost certainly be more of these polymeric compounds released into the environment, regardless of potential sinks and processes. Determining the destiny of these compounds in the environment and particularly in the ocean becomes imperative. A wide variety of organisms, including plankton and larger organisms like fish and marine mammals, can consume microplastics due to their small size. Although the precise mechanisms underlying microplastic toxicity remain poorly understood, one or more of the following three pathways will likely lead to harmful effects: ingestion-induced stress, such as energy expenditure for egestion, physical blockage, and false feeding; plasticizer leakage; and exposure to contaminants adsorbed by microplastics, such as Persistent Organic Pollutants (POPs) (Figure 3.4).

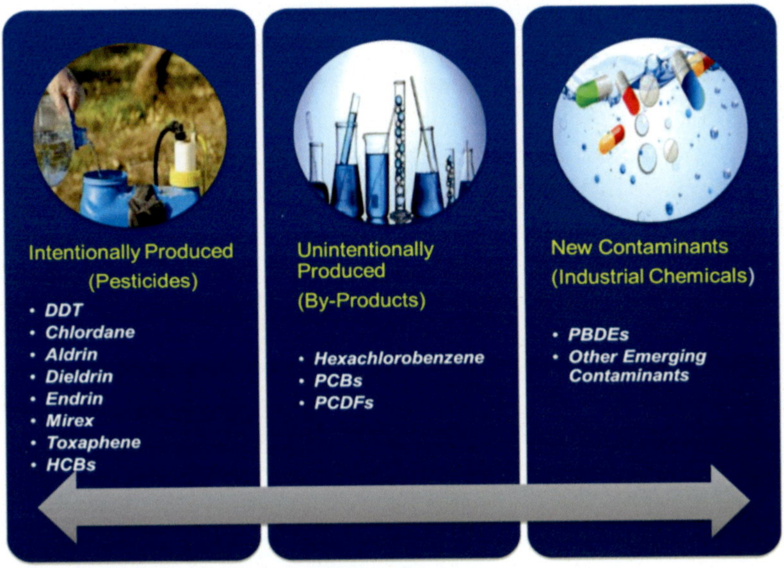

Figure 3.4. *Persistent organic pollutants.*

Source: Boulkhessaim, S., Gacem, A., Khan, S. H., Amari, A., Yadav, V. K., Harharah, H. N., Elkhaleefa, A. M., Yadav, K. K., Rather, S., Ahn, H.-J., & Jeon, B.-H. (2022). Emerging Trends in the Remediation of Persistent Organic Pollutants Using Nanomaterials and Related Processes: A Review. Nanomaterials, 12(13), Article 13. https://doi.org/10.3390/nano12132148, Creative Commons License.

3.6. CONSUMPTION OF POLYMERS AND EFFECTS

These polymers are consumed by cnidarians, rotifers, ciliates, birds, etc. Their color also affects their uptake in addition to form, size, and density. Indeed, an intriguing study conducted in the early 1990s suggested that marine species preferred to ingest yellow and tan particles, potentially as a result of mistaking them for prey. Ingestion of microplastics can directly result in internal damage and digestive tract blockage, which can lower food intake and nutrition and ultimately lead to death. For example, the entanglement of microplastic particles in gills may result in decreased respiration rates. Microplastics can be consumed in the terrestrial environment, yet there are relatively few studies on the behavior and fate of these materials in terrestrial environments, and there is, at most, scant direct and quantitative evidence of their presence in soils. However, it has been reported that livestock, specifically, sheep and goats consume plastic particles. Although no particular consequences were assessed, some authors hypothesized that these animals would experience nutritional deficits. Even though soils and aquatic habitats are very different, biota-critical characteristics like these nevertheless apply: Because they live in small bodies of water that are at or slightly below the soil's surface, many species are, in essence, aquatic. Furthermore, these substances may be consumed by mites and earthworms, which could result in their accumulation in food webs. A recent study that exposed earthworms *Lumbricus terrestris* to varying concentrations of PE particles smaller than 150 mm found significant growth rate decreases as well as increased mortality (Ju et al., 2023). Additionally, the earthworms showed size-selective egestion, as the majority of the particles they ingested were smaller than 50 mm. This could have a substantial impact on the fate and risk of microplastics in terrestrial ecosystems, as earthworms are essential to the physical characteristics of soil because of their casting and burrowing activities (Figure 3.5).

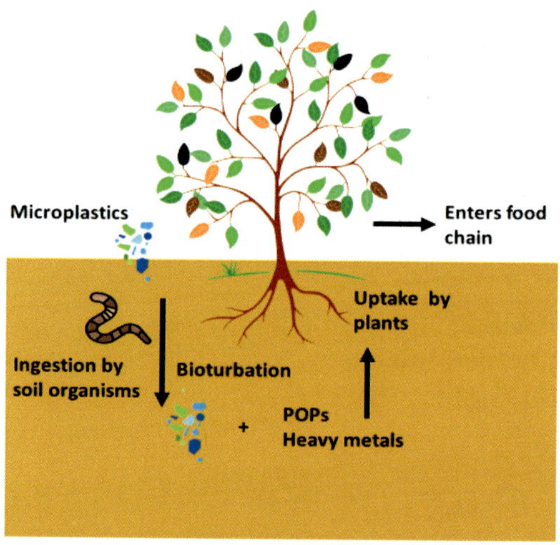

Figure 3.5. *Effects of microplastics on the soil.*

Source: https://en.wikipedia.org/wiki/Lumbricus_terrestris#/media/File:Regenwurm1.jpg, Creative Commons License.

Therefore, the presence of plastic fibers in soil for 15 years indicates that microplastics are likely to occur in terrestrial ecosystems, and their permanence has been demonstrated. However, further study is required to fully comprehend their fate and behavior. Comparatively speaking, nevertheless, plastic materials don't pose a hazard to the environment when used in isolation. Based on the growing use of submicron additives in commercial plastics, a new categorization of these compounds known as plasticides has evolved. Plastics are essentially biochemically inert materials, but the majority of them today contain additives, usually tiny molecules that aren't chemically bonded to the polymer. Moreover, partial polymerization reactions occur during the manufacture of plastic, and unreacted leftover monomers, solvents, and additives may migrate out of the synthetic matrix. As a result, the plastic materials allow them to leach. Most commonly used additives are lipophilic, meaning they may penetrate cell membranes and engage in biochemical interactions. This can result in both severe behavioral impacts and reproductive repercussions. It has been shown that polyvinyl chloride (PVC), polystyrene (PS), and polycarbonate (PC) release harmful monomers linked to the incidence of cancer and improper reproductive processes in invertebrates as well as in rats and humans. It has also been demonstrated that additives used in plastic production seep out of them and are consumed by marine life.

Furthermore, it has been demonstrated that plastics gathered from aquatic habitats also include additional pollutants, such as organic compounds that have been discovered to have adsorbed from the surrounding air. POPs, polychlorinated biphenyls, or PCBs, polycyclic aromatic hydrocarbons, or PAHs and organochlorine pesticides, like dichlorodiphenyltrichloroethane, or DDT are among these organic contaminants and should be avoided because of their strong resistance to degradation by biological, chemical, and photolytic processes.

These hazards become especially significant when dealing with microsized particles because of their high surface area-to-volume ratios. As a result, when consumed, these items may provide a fresh route for chemical exposure. It has been demonstrated that certain POPs, such as Polybrominated Diphenyl Ethers (PBDEs) and PCBs, as well as certain pesticides, mimic natural hormones and can cause reproductive abnormalities. Naturally, the characteristics of the specific polymer and the particular pollutant, such as type, density, molecular weight, and hydrophobicity, affect how quickly POPs adsorb into plastic polymers. Hydrophobic pollutants have the ability to penetrate food web/chains through ingestion and may have a role in the phenomena of bioaccumulation and bioamplification (Figure 3.6).

Despite the paucity of research on these bioaccumulative effects, certain indications appear to support these theories. For instance, it has been shown that the uptake of PS particles in human lung cancer cells is basically irreversible, and the concentration of PS particles within the cells increases linearly with time. Further research has revealed that tiny latex particles (<50 nm) can collect in the testis, liver, blood, and, possibly most concerningly, the brain of Japanese rice fish (*Oryzias latipes*). This indicates that the particles were able to pass the blood-brain barrier, emphasizing the possible risks to animal and, eventually, human health. The blood-brain barrier shields the brain from toxins and is critical for maintaining homeostasis, which is required for neural function.

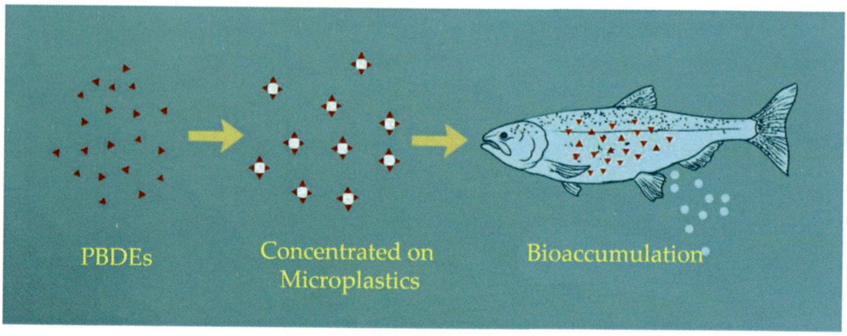

Figure 3.6. *Bioaccumulation of microplastics.*

Source: Weis, J. S., Alava, J. J. (Micro)Plastics Are Toxic Pollutants. Toxics. 2023; 11(11), 935. https://doi.org/10.3390/toxics11110935, Creative Commons License.

3.7. RESEARCH FINDINGS

The majority of research on these interactions has been done in aquatic environments. Future studies should concentrate on interactions in soils, as they have been mostly overlooked. This is because microorganisms in the soil may be exposed to organic pollutants through these pathways or other unidentified mechanisms, raising toxicity-related concerns. Other pollutants, like potentially hazardous metals and metalloids, may also adsorb at the surface of microplastics, increasing the risks that these substances may pose when released into the environment. The need for additional research on the long-term impacts of microplastic exposure is highlighted by these observed results. Microplastics from environmental samples have inherent challenges in terms of collection, handling, identification, and characterization because of their microscopic size. Therefore, it is imperative to simplify not only the nomenclature but also the methods employed to evaluate and measure the existence of these particles in obtained samples. The existing deficiency of Standard Operating Protocols (SOPs) for sampling and detection may be partially to blame for the dearth of consistent data in microplastic research. While some organizations, like the European Union, have attempted to create standardized methodologies, the majority of published reports, especially those that deal with aquatic systems, make it abundantly evident that the current efforts are insufficient to accurately sample and quantify microplastics.

The resulting varying quality of experimental data is a result of this broad range of techniques. As a result, there are evident issues with data comparability and dependability regarding the levels of microplastics in the environment. To acquire reliable and comparable data on the concentration, particle size distribution, type of polymer, and distribution across the environment and biota, it is imperative to employ standardized procedures. To create realistic and representative laboratory studies that highlight the effects of these particles on the environment and biota across all biospheres, consistent and trustworthy experimental data collection is necessary.

3.8. SAMPLING OF MICROPLASTICS

Microplastics are typically sampled using nonselective sampling methods that need to be cleaned up before being processed further in a lab. Methods that have undergone sufficient testing for their Quality Assurance and Quality Control (QA/QC) performance should be prioritized. As with other analytes, microplastics should be regarded with caution and may be over- or under-estimated depending on how the procedures are carried out to destroy or lose the samples, create more fragments, or permit contamination. We do not distinguish between primary and secondary microplastics, although pellets are typically treated separately in the literature because of their sources and relative chemical homogeneity. Given their differences in sources, environmental behavior, and ease of identification in both the environment and samples, pellets may be distinguished from fragments. They will show up in the clean sample after being blended throughout the matrix.

Samples may require a period of storage and preservation after they are received and before they can be processed in a lab. Samples can be preserved by drying (in sediments), cooling/freezing (in water; in organic-rich environments), or both. On the other hand, multipurpose samples can also be preserved in alcohol, formalin, or another reagent. It is also plausible that one of the aforementioned methods was used to fix and then preserve the large animals in question when the stomach contents of those organisms were analyzed. Although it is not anticipated that preservation will significantly change microplastics, it may cause some of their characteristics, most notably color to disappear. Additionally, samples must be dried at relatively low temperatures because temperatures exceeding 40°C can mischaracterize fibers. Before being subjected to optical or instrumental examination, microplastic samples taken from marine environments are often filtered to provide dry samples that enable particle isolation. It may be necessary to resolve the complex matrix present in the filtrate to confirm and quantify the presence of microplastics. Since most approaches change the microplastics themselves and degrade their physical and chemical characterization, removing these matrices has proven to be a significant difficulty. Prior to conducting any characterization, it is crucial to remove organic debris from plankton or stomach contents in order to isolate microplastics from the sample. The matrix contains essential information that can aid in the understanding and contextualization of the microplastics, therefore it shouldn't be disregarded. For ecological interpretations, a biogeochemical characterization of these matrixes is equally critical. Since the matrix removal techniques are destructive, extra samples or subsamples should be obtained in order to perform additional analyses that enable the study of the plankton community and other significant biogeochemical variables that could aid in the explanation of processes involving microplastics at sea. Whenever possible, it can be a good idea to store away extra samples as a backup in case of contamination.

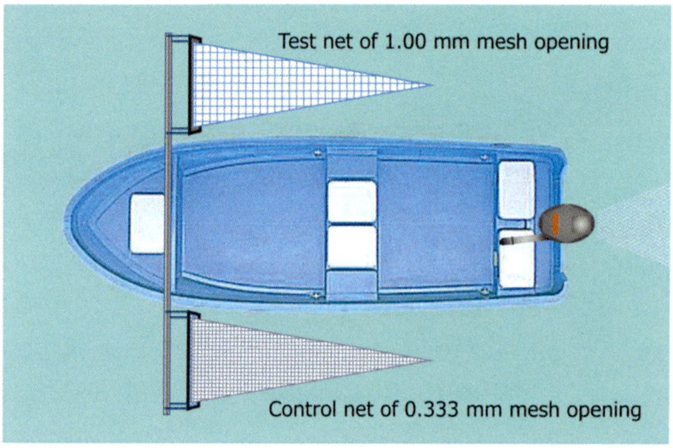

Figure 3.7. *Neuston tow nets.*

Source: Tokai, T., Uchida, K., Kuroda, M., & Isobe, A. (2021). Mesh selectivity of neuston nets for microplastics. Marine Pollution Bulletin, 165, 112111. https://doi.org/10.1016/j.marpolbul.2021.112111, Creative Commons License.

Sample storage should always be taken into consideration for future use at the conclusion of a project. Sample contamination is a notion that marine chemists and geochemists have been familiar with for decades (Johnson et al., 2008). For a very long time, the majority of samples were collected with little to no consideration for this crucial component, and microplastics were not easily recognized as a passive source of sample contamination. The prevalence of synthetic polymers in the environment may have an impact on microplastic sample handling and instrumental analysis. As a result of the recent drop in particle size and the awareness that this analyte is present almost everywhere, preventing contamination has become a crucial component of the planning process for every sample operation. It is crucial to minimize the use of plastic in synthetic clothes, field gear, and labware whenever feasible. Fibers can arise from a variety of sources, including lab coats and equipment, and they pose a risk of contaminating samples. At this stage, neglecting contamination control could compromise the attempt to ensure that samples have arrived safely from the field in the lab, and vice versa. Seawater sampling is perhaps the most widely used technique for studying microplastics. It is also one that easily includes the media and the general public, which makes it essential for scientific communication regarding the issue. It frequently involves trips to the high seas, and while it can be conducted from inflatable boats and oceanographic vessels of various sizes, sail boats and tall ships are the most visible to various societal segments. In actuality, a sizable number of surveys are conducted during circumnavigation trips that are organized by NGOs with varying degrees of scientific skill and national representation.

3.9. COLLECTION DETAILS

Samples must be filtered or sieved to segregate microplastics into a solid phase (Nguyen et al., 2019). Cleaning procedures that differ in quantity and type based on the type of plankton present in the sample are necessary to provide a distinct separation from other particles. Before samples are transferred and undergo additional processing in a lab, microplastics can be isolated using techniques like peroxide oxidation. The literature also reports approach combinations and changes based on the requirements of each project. It is crucial to keep in mind that the water may contain a variety of organic materials from plankton as well as suspended particles from river flow, and an abundance of effluents. Consequently, it may be necessary to combine the cleaning procedure with thick mineral matrixes. Neuston tow nets (Figure 3.7) deployed in Pacific Central Gyre waters were the first to reveal the widespread dispersal of microplastic pollution on the sea. Subsequent research using the same methodology revealed that this pollution may exist in all aquatic environments on Earth. Initially, it was assumed that the majority of microplastics would be found in the upper centimeters of the water column since they are typically less dense than saltwater.

The most popular ones for high seas are the catamaran and manta trawls. Tows typically take between one and five knots and can take anything from minutes to hours. In order to provide integration of the sampled area, Neuston nets must always be fitted with flowmeters and have the tow distance recorded. Manta nets come in several designs and functionalities; some can be taken apart for portability. But this contaminant's actual polluting nature only becomes apparent when it interacts with the aquatic/marine biota and causes damage that may be detected and occasionally measured. This relationship must be considered during sampling and the sample management process that follows. When researchers arrange their collection and data treatment, they must consider the problem of demonstrating whether, how, and to what extent microplastics can impair biological systems. The research has documented toxicological experiments, which illustrate that the biota is truly at risk from interactions such as intake and desorption of contaminants. The primary issue is with field research that attempts to establish a connection between microplastics and the biota as food chains, particularly in ecological environments. Due to the possibility of transmission across marine food chains, the adsorption and deposition of persistent organic contaminants in microplastics posed an ecological risk for a while. However, it is currently evident that human populations are similarly vulnerable to the risks posed by seafood and other sources, given their essential role in aquatic food chains

3.10. PROBLEMS WITH MICROPLASTICS

In recent times, investigations on microplastics have found that seafood is one of the most significant target matrices. The discovery of micro and nanoplastics in seafood and other food items is concerning, but it should not be interpreted as a surprise but rather as confirmation. However, the discovery of microplastics in beer and honey came as a surprise. It is well known that microplastics get integrated into seafood and end up in the muscle tissue after passing through the digestive system. They still need to be studied to see if humans can integrate them.

The biggest threat to people right now is posed by bivalves and other filter feeders, which is why they will likely soon be subject to mandatory monitoring for the presence of contaminants. Techniques for extraction, preservation, and sampling have thus been devised to ensure the accuracy of the data that will be made available to the general public. Microplastics are ubiquitous in aquatic habitats and have also migrated through the biota (Anbumani and Kakkar, 2018). The primary method for contamination is by food, while respiration has also been mentioned as a possible conduit. The complexity of microplastics which can mix with fragments of other prey, sediments, and indistinguishable organic and inorganic materials as well as their occasionally very small mass for small fish, for example, make it difficult to identify the types and quantities of microplastics in dietary matrixes. But the most crucial factor could not be the accuracy of microplastic detection, which is generally undervalued, but rather the sample design and methods used for animal sampling. The animal sample needs to make ecological sense and be related to the sources, distribution mechanisms, and spatiotemporal density of microplastics that are expected to vary in the environment. Animals can be dissected and their stomach contents examined to identify the most essential dietary items, which may include microplastics, once a reliable ecological interpretation has been obtained. Thus, additional details regarding this contaminant's environmental pathways can be provided, and it can be taken into account for statistical treatments in the future as a food item (Figure 3.8). This would be expected in the case of apex predators, who aren't filter feeders and should therefore be contaminated up the food chain.

This illustrates yet another way that microplastics are ubiquitous in aquatic environments. Visual searches on the contents of the stomach and intestines can be conducted directly with the use of optical instruments, with or without prior sifting, acid/basic digestion, or elutriation. A stereomicroscope, calipers, and individual item washing could be used to visually separate microplastics from stomach and/or intestine digests. It is also feasible to combine sample cleaning and detection techniques.

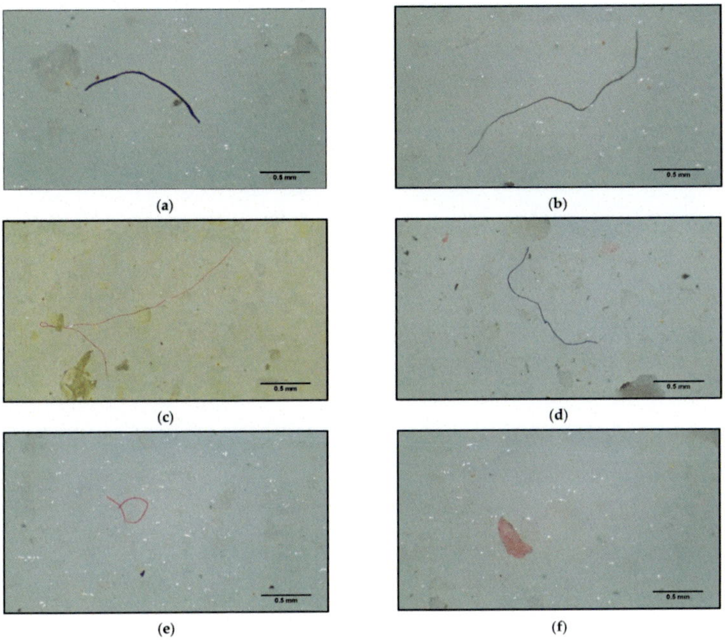

Figure 3.8. *Occurrence of each type of microplastics in the king mackerel (S. guttatus) from the lower Meghna estuary in Bangladesh examined under a stereomicroscope.*

Source: Hossain MB, Pingki FH, Azad MAS, Nur A-AU, Banik P, Paray BA, Arai T, Yu J. Microplastics in Different Tissues of a Commonly Consumed Fish, Scomberomorus guttatus, from a Large Subtropical Estuary: Accumulation, Characterization, and Contamination Assessment. *Biology*. 2023; 12(11):1422. https://doi.org/10.3390/biology12111422, Creative Commons License.

One effect of this kind of pollution that has long been documented is the ingestion of microplastics by freshwater and marine biota. It is only recently, nevertheless, that it is likely the case that entire food chains are affected by the occurrence. Microplastics have been identified in the digestive tracts of fish of ecological, social, and commercial significance in certain ecosystems. This indicates that the pollutant is widely present in the fish's feeding grounds. It is crucial to note that not all research has been able to demonstrate the bodily damage that eating microplastics by wild animals might cause. It is also critical to consider whether the damage observed affected an individual, a community, or a sub-organism. The latter is most likely the most difficult task in ecology and the management of living resources. Particles with varying sizes, shapes, colors, and compositions are known as microplastics. As such, they provide a range of substrates on which bacteria can colonize the aquatic environment and its many compartments. There are reports that this new artificial environment, which is known as the Plastisphere, can be thought of as a compartment itself because plastics have been found in the water, sand, and to some extent, the biota. Studying this specialized microbial community begins with accurate and deliberate microplastic sampling. Any harsh/rough treatment such as hydrogen peroxide, drying, and acid/base solutions would quickly wipe off the population that is being studied. Microplastics and microbiological approaches should be combined in sample preservation and treatment methods (Wang and Wang, 2018). It is crucial to give this matter more thought as ingesting microplastics contaminated with viruses can pose a threat to biodiversity, particularly for species that are ecologically vulnerable or live near the coast. Additionally, opportunistic organisms may travel and spread farther than anticipated geographically in conjunction with microplastics in addition to larger objects.

3.11 ASSESSING MICROPLASTICS IN AQUATIC ENVIRONMENTS

Today's society uses an unparalleled range of polymers, which results in an equally wide range of wastes and secondary microplastics in aquatic habitats. Higher densities may gradually sink to the bottom where they will eventually aggregate, whereas less dense materials prefer to either float or deposit on beach surfaces. Their density can also be altered and changed by aging and microbial fouling, which aids in deciding their eventual fate. The same is true for primary microplastics, albeit due to their distinct forms and surface textures, they can be slightly more noticeable in the environment (pellets) and easier to identify when inspected visually, with or without assistance. The closest and easiest-to-access microplastics for the average person are those discovered on the beach. In fact, the science behind it all began with the study of microplastics left behind on beaches, and a variety of techniques have been employed to collect and analyze sediment samples in the lab. The two most popular methods for separating microplastics from the matrix are hydrogen peroxide oxidation and saline solution flotation, which is a density-based separation technique. As long as the decisions are compatible and prevent the loss of microplastics in both quantity and quality, combinations of physical and chemical procedures are also feasible. Since beaches are three-dimensional sedimentary structures, sampling from them must take into account that various ongoing depositional temporal and spatial scales will be reflected in each habitat. Planning a microplastics sample project may benefit from taking into account ecoclines. Whatever the situation, sampling must always be expressed as a volume (m^3) or an area (cm^2) of sediments. Comparing reports of microplastics in freshwater or the ocean, units can be difficult to compare and vary greatly. The units that can be used to write the report will also depend on the sampling settings chosen.

Plastic materials should be avoided while using sampling equipment, which can range from basic to advanced. This is especially true if analysis for persistent organic pollutants will be performed on the pieces or pellets at a later time. Glass vials that have been decontaminated would be perfect, however, their handling, cost, and storage space are limitations. Digging trenches to collect microplastic samples from beaches can provide information on the seasonal and interannual deposition of primary and secondary microplastics. Additionally, it can aid in the investigation of the contamination of the beach's deeper layers with these chemicals, which might not be present if they didn't have the plastic support to keep them buried. This is in addition to the analysis of petroleum hydrocarbons and persistent organic pollutants. As depth increases, subtidal sampling becomes progressively rarer and may be the most challenging. Since movement may be involved, the lack of this knowledge makes it more difficult to comprehend the processes that move microplastics to the ocean and deeper sediments. It can take a long time and be challenging to separate from the matrix. It can be accomplished by a sequence of processes that always involve a combination of chemical (oxidation/reduction) and physical (density-based) steps. Recently, specialized hardware was created to make the task easier. Beyond just removing the microplastics from the bulk sample, it is critical to determine whether the work put into separate size classes as well as shape, color, material, and other factors among the resulting microplastic extract will be valuable when it comes time for statistical analysis. Other factors such as grain size, carbonate contents, beach profile, wave climate, beach use, and so forth should be taken into account while studying intertidal sediments on beaches. Studies on microplastics should ideally be coordinated with studies on beach ecology, particularly those that look into the food ecology of scavengers (Fanini et al., 2020).

Since access to beaches can be expensive, what we refer to as conventional microplastics sampling from beach sand can be time-consuming and challenging to deploy at large scales. An alternate scanning method for large-scale surveys has been proposed, which involves counting surface microplastics on beach sand by photographic means and on-site. While counting microplastics on the spot can be a difficult undertaking, it is occasionally the only option available, such as when it is not feasible to get all of the samples back to the lab. Among these are the calibration and quality assurance of the observers vision. Once calibrated to ensure a reasonable correlation with conventionally examined samples, photographs can be digitally evaluated and used as a preliminary evaluation to identify monitoring sites. The amount of microplastics in beach sediments determined by image analysis will always be underestimated since buried pieces will not be taken into consideration. However, these interpretations may be helpful, for example, in the identification and semi quantification of pellets. Because of the preventative actions implemented at their maritime and continental sources, it is anticipated that the amount of virgin plastic resin pellets in the marine environment will gradually decline. Therefore, this alternate sample strategy could aid in verifying if sources are not just migrating southward to emerging nations and in testing this hypothesis on a broader geographic scale. Above all, though, taking pictures with cellphones can simultaneously engage citizen scientists in the fight against microplastic contamination in coastal and marine habitats and provide high-quality research. Virtually every environmental compartment where microplastics have been searched for has yielded results. Nonetheless, the majority of studies take into account static, snapshot conditions and are more effective in providing answers to concerns about the presence or absence of microplastics and their properties. Papers that examine and validate mechanisms related to microplastics are not often published. If microplastics may truly sink to the ocean floor and become part of deep-sea sediments, then it is one of the most significant unanswered concerns in the field. Despite being discovered in the deep water, there is still good reason to be skeptical about the exact mechanism and rate of sinking. To determine the length of time microplastics spend in seawater and whether sinking is an important removal method, such rates would be required.

Deploying layered sediment traps at arrays of moored buoys used for meteo-oceanographic and climate investigations is one potential sampling approach to investigate the sinking process of microplastics into the deep sea. Commercially available sediment traps for marine snow that can be used for microplastic sampling increase the likelihood of a timely and effective deployment and resource optimization. In the near future, it is anticipated that its implementation at various locations can produce more consistent responses to these queries. Some have claimed that collecting samples for microplastics in this manner will make it more difficult to use the samples for other purposes (biogeochemical analysis). For marine scientists who are interested in both kinds of factors, the solution to this puzzle is still a difficult. Visual identification enables us to quickly sweep over MPs, which is frequently sufficient to distinguish between plastic particles ranging in size from 0.25–5 mm in the bulk sample. This is a quick and inexpensive method, but it has many drawbacks because we cannot distinguish plastic particles from natural soil/sediment components and cannot identify MPs smaller than 0.25 mm. As a result, it is common to classify particles incorrectly that have the same morphology and/or color as MPs. Even seasoned observers typically find fewer than 100% of all plastic particles in a sediment sample. The researchers also postulate the degree of variability in detection varies greatly between types, colors, and observers. Dissection and visual inspection of the digestive tract of certain animals (fish and lobsters) have been considered adequate techniques for biological samples. However, they are ineffective for many other species that eat MPs with smaller sizes. Before a visual or microscopical identification of MPs with sizes not apparent to the unaided eye, various procedures have been attempted, including the digestion of biological material (tissues and fecal pellets) to eliminate organic debris. Nevertheless, it is also critical to consider the possibility that some of these techniques could lead to misidentification because some of the chemicals employed can alter the morphological traits, like the size and/or color of microplastics. For instance, researchers demonstrated that H_2O_2 is ineffective in removing biological material and may also drastically shrink MPs (<1 mm) in size (Zhang and Chen, 2020). Alkaline and acid digestive techniques, on the other hand, may also encourage MP discoloration. The amount, size, and form of MPs consumed by various organisms have also been examined visually using light microscopy examination of fecal pellets.

3.12. IDENTIFICATION REPORTS OF MICROPLASTICS

The combination of optical microscopy and visual identification is a commonly used technique for various environmental samples due to its low cost, simple criteria, and ability to be conducted in situ. However, it can be very challenging to verify that the particles found are actually MPs in some cases without the use of complementary spectroscopic techniques like Raman spectroscopy or Fourier Transform Infrared Spectroscopy (FT-IR). This is especially the case for particles with sizes smaller than 1 mm. After analyzing samples from Surface Microlayer Water (SML) and Beach Sediment (BS) using FT-IR and optical microscopy, researchers found that when the latter technique was employed for identification, there was a greater abundance of MP fragments for size classes <500 mm. When form categories were taken into consideration, this tendency did not hold true because there were fewer fibers in both types of samples (SML and BS) when they were analyzed using FT-IR. Expanded Polystyrene (EPS) and fragments showed the opposite tendency, indicating that the identification of MPs may be overstated or underestimated based on both size and shape. However, researchers advise against using FT-IR and Raman spectroscopy methods unless a small number of samples need to be analyzed due to the skill and time necessary. Both approaches might be applied to a large number of samples, beginning with the FT-IR screening of MPs in a small number of samples to determine which MP groups are the most pertinent and typical. Then, visual identification is supported by this initial characterization of MPs. In response, scientists found that 63% and 83% of MPs <50 mm and MPs >100 mm could be identified correctly when comparing visual identification by optical microscopy with identification done by the Raman microspectroscopic technique. Fibers of all colors were the MPs with the highest recognition success rate when forms were taken into consideration. These scientists claim that this occurred as a result of the structural features for identification.

Using tweezers, a heated needle was used to make contact with the particles. The particle is identified as a plastic polymer if a needle makes it stick and leave a mark. Nevertheless, it remains uncertain to what range of MP sizes this method may be applied for a dependable identification. The shapes and sizes of MPs that are found in the environment can vary greatly. These variations are caused by three factors: the initial form of the primary plastics; the process of degradation and erosion of the particle surface (biological breakdown, photodegradation, physical forces, etc.,); and the residence time in the environment. Sharp edges that indicate a recent introduction and smooth edges linked to a long residence time are the causes of each fragment's shape. As a result, it appeared that these categories were determined case-by-case following an earlier visual examination to determine the primary forms seen in the samples. MPs can take on a variety of forms in the environment, and it's possible that some of these forms could seriously injure host organisms physically. But this hasn't been shown yet, at least not for MPs. Certain species appeared to be discriminating feeders, both in terms of shape and size. Large particles are rejected by *Arenicola marina* L., which exhibits size-based selectivity.

The distribution of shapes in the medium most likely determines the form of the most absorbed particles, even though there may be some selectivity in terms of size. When researchers examined the stomach contents of several mesopelagic and epipelagic fish species, they discovered that fragments accounted for 94% of the contents, with films and fishing lines coming in second and third, respectively. When non-ingested MPs were gathered from neuston trawls, comparable percentages were noted. In accordance with their content in the sediment, the distribution of MP sizes and shapes consumed by holothurians was rather near to expectations. Despite the appearance of MPs being absorbed passively in all forms, certain MP shapes may inhibit egestion after ingestion. These researchers demonstrated the presence of balls of strands that were not removed by the digestive process in the stomach of *Nephrops norvegicus* L.

as a result of stomach activity. But now that size is taken into consideration, selectivity could not occur during ingestion; nonetheless, smaller sizes may favor MP accumulation in tissues, as shown in the case of mussels. Several studies have also used their morphology to infer information about their paths and origins. For instance, fibers were the most common form of MPs discovered in harbors, beach sediment from Belgium's coastal region, neustonic samples taken at the Black Sea, agricultural soils containing sewage sludge inputs, and atmospheric fallouts gathered from both dense and less densely populated urban areas in the Paris region (France). Because of their similarities, the microbeads discovered in Hong Kong's surface coastal water were linked to those found in personal care products. The researchers claim that these beads entered the sea through discharges from nearby treatment facilities or from wastewater that washed into the rivers untreated.

Chapter 4

EFFECT OF MICROPLASTICS IN MARINE ECOSYSTEMS

Contents

Introduction........76
4.1. Sustainability of Resins........76
4.2. Impact of Plastics........77
4.3. Pollution Due to Plastics and Additives........79
4.4. Plasticizers........80
4.5. Health Effects of Exposure........81
4.6. Pigments........83
4.7. Polyolefins........84
4.8. Packaging........86
4.9. Organotins........87
4.10. Codex Alimentarius........88
4.11. Chemicals........89
4.12. Bottle Materials........90
4.13. More on BPA........91
4.14. Research Findings........92
4.15. Antimony in Bottled Water........94
4.16. Floating Plastics........96

INTRODUCTION

Plastics are being produced more widely worldwide, and plastic-based goods are now an indispensable component of daily life. In just 75 years or so, plastics have successfully replaced traditional materials in a wide range of application sectors. The possible toxicity of plastics and rubber materials to the environment, especially to humans, is a crucial component of their sustainability. On the other hand, plastic resins are harmless substances made of macromolecules that are neither absorbed or digested in the digestive tract. The various additives put into the resin during product processing and manufacturing, the leftover monomers or catalysts stuck in the resin, and the chemicals the plastics absorb from the environment are all thought to be responsible for the frequently seen toxicity of plastics. The additives are the ones found in plastic products in the highest concentrations.

4.1. SUSTAINABILITY OF RESINS

The sustainability of plastics depends on understanding and assessing the frequently used chemicals, their possible toxicity, and the methods by which they are ingested by human users. Global output of plastic resin reached 322 million metric tonnes (MMT) in 2015, maintaining a generally steady upward trend in production since the material's initial commercial introduction in the 1950s (Idumah et al., 2019). Assuming present per capita consumption rates, the robust increase in demand for plastics has exceeded that which can be ascribed to population growth. This points to a global trend of rising resin use per person, particularly in the wealthier regions. Asia/Pacific nations dominate the world resin supply, producing half of the resin produced worldwide. Together, the USA and Western Europe produce only 15–17% of the world's resin. It is easy to understand why plastics have become so popular: they offer a low-cost, lightweight (low density), biologically inert, incredibly durable, and thermally formable into intricate patterns. Of the fossil fuel that is extracted for usage, only around 4% is used as the feedstock for plastics. Plastics need less energy to manufacture and recycle than metal or glass, which lowers carbon emissions. Plastic resin manufacturing technology has developed over the years into effective, low-waste, low-emission processes. Without a doubt, the demand for plastics will rise for the foreseeable future given the worldwide industrial network supporting this growth.

The industry's environmental sustainability has been questioned in spite of its extraordinary success and current strong status. To secure the survival and well-being of both the present and future generations, sustainable growth in any industry necessitates the responsible and waste-free use of resources as well as the preservation of the natural capital that is essential to life on Earth. The social cost of a technology in terms of energy use, material requirements, and environmental impact must be weighed against the social benefits of the technology in order to determine its environmental sustainability. The significant increase in plastic consumption highlights the material's numerous advantages for society. A confusing array of plastics is seen on a daily basis in clothing, coatings, furniture, cars, buildings, and packaging. Their extensive use in packaging prevents food waste while keeping food and medication from spoiling, and they save a substantial amount of energy when used in building transportation applications. Plastics have a direct financial impact on the production of energy through applications like solar panels and wind energy installations. There is a compelling argument to be made for the delivery of services across numerous application sectors using only 4% of petrochemical feedstock. Fossil fuel feedstocks are finite resources that are quickly running out. This need for feedstock can be somewhat reduced by raising material quantities or chemically recovering polymers. In any case, the industry's dependency on non-renewable feedstock will decrease due to the increased interest in alternative plant-based feedstock for the synthesis of plastic resins.

4.2. IMPACT OF PLASTICS

The expense of externalities connected with plastics, particularly those that may have an impact on the environment or public health, is more difficult to justify. These externalities, particularly those compounds in plastic resins and goods that could be hazardous, must be phased out in order for plastics to flourish sustainably. Only a small number of fifty chemically different groups of plastics are employed in high-volume applications. These include polyethylene terephthalate (PET), polyvinyl chloride (PVC), polystyrene (PS), polypropylene (PP), and polyethylenes (PE). The most widely used plastic in the world, PE, come in a variety of grades with marginally varied bulk density ranges, including high-density polyethylene (HDPE), low-density polyethylene (LDPE), and a Linear low-density polyethylene (LLDPE). Plastic packaging frequently uses commodity polymers; PET, for example, is frequently utilized in soda bottle applications. About 40% of thermoplastic production in the USA is employed in packaging, with the remaining 25% going toward the manufacture of consumer goods and only 20% going toward building construction (Jagadeesh et al., 2022). PVC and PE are the most often utilized plastics in building construction, whilst polyolefins are the plastics used in packaging.

To adjust the qualities (plastic) or regulate degradation during thermal processing, chemicals are usually added to plastic resins before they are used to fabricate goods. Before the molten plastic is thermally formed into the finished items, these additives or chemicals are thoroughly blended with the resin. Global additive output reached 12.6 MMT in 2013 and increased at a rate of about 4.8% per year over the previous five years. Certain additives, like flame retardants, are anticipated to expand at a significantly faster rate. The plastics that require the greatest amount of additives are the plasticized PVC composites. Pure plastic resins are safe, but the additives that seep out of plastic items raise questions about their potential toxicity. Some of the monomers, dimers, trimers, or oligomers that are held within the solid resin with PS, PET, and polycarbonate may also leak out. The possible risks to the environment and/or public health associated with plastic goods are caused by the presence of these additives. When food comes into contact with packaging polymers that include hydrophobic chemicals, the additives may seep into the food, particularly in foods that have a high lipid content. This provides a pathway for the ingestion of these chemicals by humans. This is an even more serious concern when it comes to the additive chemicals' potential for bioaccumulation in the individual.

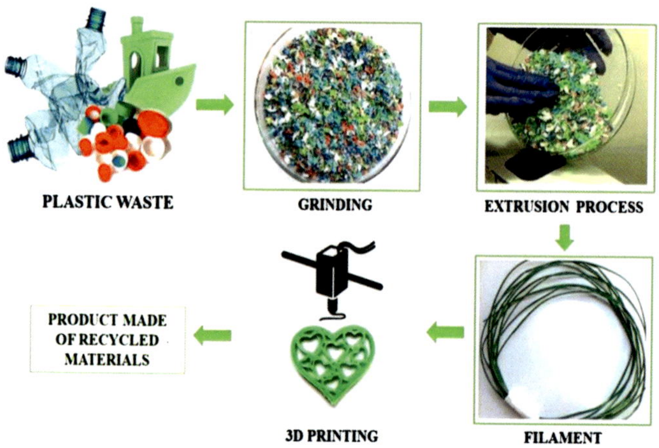

Figure 4.1. *Medical plastics.*

Source: Selvaraj, S., Prasadh, S., Fuloria, S., Subramaniyan, V., Sekar, M., Ahmed, A. M., Bouallegue, B., Hari Kumar, D., Sharma, V. K., Maziz, M. N. H., Sathasivam, K. V., Meenakshi, D. U., & Fuloria, N. K. (2022). COVID-19 Biomedical Plastics Wastes—Challenges and Strategies for Curbing the Environmental Disaster. Sustainability, 14(11), Article 11. https://doi.org/10.3390/su14116466, Creative Commons License.

Relevant products include the plastic films that cover the interior of paperboard packaging, the epoxy lining found inside tin cans, various types of plastic bottles, and plastic food packaging. The majority of packaged foods have extensive contact times, and even in cases where the kinetics of leaching are slow, considerable migration may still take place. The degree of migration and, consequently, the toxicity presented by additives, are generally determined by the temperature, pH, and hydrophobicity of the food. A comparable route for leached additive intake is also presented by newborns playing with plastic toys. This problem has been somewhat resolved by recent regulatory restrictions on the use of certain phthalate plasticizers in toys. The primary method of exposure is ingestion. Important exposure pathways include inhaling volatile chemicals from car interiors and coming into contact with bodily fluids through medical plastics (Figure 4.1).

4.3. POLLUTION DUE TO PLASTICS AND ADDITIVES

The potential for microplastic litter to concentrate persistent pollutants in saltwater is a growing issue for the marine ecosystem. Marine organisms have the ability to consume these plastics, which allows them to enter the marine food web. It is recognized that a variety of marine animals, such as fish, birds, turtles, invertebrate species, and marine mammals, consume plastics. At least 600 species are known to be impacted by plastic waste at sea, with the majority of these being affected via ingestion (Ryan, 2019). The ensuing toxicity could be substantial if the chemicals in them are accessible to the organism. When additives go up and down the trophic levels of the marine food chain, they can become biomagnified.

The pH, surfactants, hydrophobicity, and length of time the plastic remains in the organism all affect the net entry of additives through this route. High levels of lipid gut contents naturally increase the absorption of hydrophobic chemicals like the phthalates in PVC. If the additives are Endocrine Disruptor Chemicals (EDCs), the threat to the ingesting organism is even greater because even at very low dosages, well below the conventionally accepted toxicity levels, these chemicals can markedly interfere with the physiology of the organism, with long-term adverse effects.

Reproduction, behavior, metabolism, and fetal development are all influenced by the endocrine system. It's a sophisticated system that needs synchronized hormone signals working together to control physiology. EDC chemicals cause sickness by either mimicking or interfering with hormones in the body. Since the endocrine system secretes hormones that typically act at extremely low concentrations, EDCs likewise function at very low doses. This presents the primary challenge in determining the toxicity of this group of compounds. When adjusting human physiology, their dose-response curves are nonlinear and frequently U-shaped or inverted U-shaped. As a result, EDCs may have little to no effect at greater concentrations but may cause physiological reactions much below the reference or safe levels determined by toxicological investigations. As a result, EDCs may not have a harmful effect when tested for toxicity using traditional screening methods. Another issue is that the effects are frequently long-lasting; in animal models, animals exposed to EDCs in one generation may have negative effects on animals in subsequent generations.

4.4. PLASTICIZERS

Typically, plasticizers are low-molecular-weight liquids that are used with plastic to soften the material and increase flexibility (Godwin, 2017). For example, plasticized PVC is a flexible that finds application in many household items such as shower curtains, membrane roofing, car seat covers, blood bags, and medical tubing. When PVC is plasticized, it changes significantly from unplasticized or rigid PVC in terms of its mechanical characteristics and feel. The latter is commonly found in window frames and PVC pipes. Phthalates are commonly employed in plasticized PVC; in soft PVC formulations, plasticizer levels as high as 60–70% are not uncommon. Since they are not covalently bonded to the macromolecular chains, the plasticizer molecules are dissolved in the polymer matrix and are free to move around inside the resin. Although the precise mechanism by which plasticizers work is not entirely known, it is thought that suitable plasticizer molecules form a close association with the polymer chains, adding extra free volume to the bulk plastic matrix and causing it to soften. The drop in plasticized polymers' glass transition temperature Tg (C), which is inversely correlated with the fractional free volume of the polymer, lends credibility to this argument. However, this mechanism by itself frequently falls short of providing a complete explanation for plasticization in all polymer systems. Due to the widespread usage of phthalate plasticizers, plasticizers currently make up the greatest portion of the worldwide additives industry. At the moment, they make up almost half of all additive sales worldwide. The PVC industry uses the majority (94%) of the approximately 6 million tonnes of phthalate plasticizers (Wang et al., 2018). In PVC formulations, di (2-ethyl hexyl) phthalate is the most commonly used plasticizer.

High levels of plasticizers in plastic goods, and consequently in plastic waste, pose a threat to the environment because they make the chemical species more accessible to organisms that consume plastics, particularly those found in maritime environments. There is a possibility that ingesting species with the capacity for bio-magnification at higher trophic levels of the food pyramid will bio-accumulate phthalates with a higher octanol-water partition coefficient (Kow). Biomagnification, however, might not always result from bioaccumulation. For example, DEHP does bioaccumulate but is not highly biomagnified since it is efficiently metabolized by species further up the food chain. When ingested by small animals, like copepods, the body concentrations can be rather high. The majority of species' capacity for bioaccumulation, however, is still unknown. Although the PVC business mostly uses phthalates, it also uses other kinds of plasticizers, like epoxidized vegetable oils. Low-molecular-weight liquid polyesters and alkyl esters of dicarboxylic acids (adipic, azelaic, and sebacic acids) are also utilized as plasticizers. Food and water are the main ways that humans are exposed to phthalates, with plasticizers being ingested from packaging or the surrounding environment. Foods vary greatly in their concentrations, with foods high in fat having comparatively greater levels of phthalates. For example, high levels of some phthalates have been recorded in butter, peanut butter, and milk. Naturally, food packaging can have high concentrations, some of which may leak out based on the contents of the package.

4.5. HEALTH EFFECTS OF EXPOSURE

Many harmful health effects can arise from phthalate exposure (Kamrin, 2009). Though other impacts have also been observed, these are mostly linked to male reproductive development and function (poor sperm quality, higher sperm DNA damage. The USA, EU, and Japan have recently placed regulations on the use of certain phthalates, particularly in toys, as a result of these uncertainties/concerns. Emerging, potentially green, plasticizers such as bis (2-propylheptyl) phthalate (DPHP) or bis(2-ethylhexyl) adipate (DEHA) are less hazardous substitutes for traditional phthalates. Ensuring their respective merits concerning the toxicity of their breakdown products and their leachability from plastic matrices is crucial.

These are additives made of organo-bromine that are added to polymer formulations to regulate the end product's ignitability. Due to their relatively high concentrations (10–20%) in flame retardant compounds and the fact that some of them are endocrine disruptor chemicals, they are of significant environmental significance. They disintegrate or get suspended in the plastic and have the same potential to leak into the environment as plasticizers. Flame Retardants (FR), which include polybrominated diphenyl ethers (PBDEs), are the class of plastic additives with the fastest rate of growth because of their widespread use in the textile, transportation, and building industries. PBDEs are used as fire retardants in resins PE, high-impact polystyrene (HIPS), ethylenevinyl acetate copolymer (EVA), and acrylonitrile-butadiene-styrene COPOLYMER (ABS). It is also found in polyurethane foam, textiles, and plastic home items as TV/computer housing (Figure 4.2).

Figure 4.2. *Computer housing.*

Source: https://en.wikipedia.org/wiki/Computer_case#/media/File:Gabinete99.jpg, Creative Commons License.

Because the fire-retardant mechanism depends on bromine being available, the deca-bromo congener works especially well. In the event of a fire, the PBDE thermally dissociates into Br radicals, which then create HBr by removing hydrogen from the polymer. Less reactive compounds are produced when this HBR reacts with the gaseous byproducts of polymer pyrolysis and oxidation that are created when the plastic burns. Commercial PBDE additives are congener combinations, also known as technical mixtures. For example, the widely used penta-BDE additive, which is produced in the USA, is a combination of tetra-, penta-, and hexacongeners. Since 2003, the European Union has prohibited the use of PBDEs, and the United States of America has stopped producing penta- and octa-PBDEs (Zhang et al., 2014). The industry is also voluntarily phased out of the deca-bromo congener. The manufacture of BDE congeners containing 4–7 bromines was limited by the 2009 Stockholm Convention. Nonetheless, a small amount of PBDEs produced in Asia may still make it into the USA and Europe. Despite being no longer produced in the majority of the world, PBDEs that were once employed by the industry still exist in the environment. Although PBDEs are categorized as Persistent Organic Pollutants (POPs), they are readily photolyzed outside. Long-term studies have revealed that the additives are present in a range of environmental contexts, including milk and human plasma. For example, it was observed that for people living in the USA and Canada, human milk contained about 20 μg/kg of PBDE. The PBDEs are thought to be Endocrine Disruptor Chemicals (EDCs), in contrast to the majority of other additives. Moreover, the presence of additional active chemicals in the system significantly alters the action of EDCs, including PBDEs. The use of PBDEs in plastics has been severely restricted as a result of a combination of these concerns. There are already other flame retardants on the market, such as alkali metal oxides; at the moment, phosphorus and aluminum hydroxide-based alternatives dominate the market.

4.6. PIGMENTS

Inorganic pigments or dyes are used to color plastic items. Oxides, sulfides, chromates, and complexes based on metals including cadmium, zinc, nickel, antimony, manganese, cobalt, aluminum, iron, copper, and molybdenum are among the often-used inorganic pigments. The most common white pigment is titanium dioxide, often known as rutile, and is mostly utilized in rigid PVC products. Carbon black pigment is typically used to provide a black color. The uniformity of color of a product is crucial for consumer acceptance and is mostly determined by the pigment's dispersion in plastic as tiny particles (0.01–0.1 μm). The colored plastic portion becomes opaque due to the scattering and absorption of light by pigment particles. Organic dyes are soluble in the polymer and solely absorb light they do not disperse. Consequently, the plastic tends to be transparent or translucent even at high dye concentrations. There is less information available on the possible toxicity of colorants in plastics than there is for textile dyes. Since lead and cadmium-based pigments were phased out a while back, colorants have been utilized without raising any serious concerns. However, there are sporadic reports of possible toxicity, such as those involving the white pigment titania. The particles of inorganic pigments are typically too large to go to the plastic product's surface and disperse into the surrounding environment. Nevertheless, surface layer ablation that releases pigment over an extended period of time can still cause this. The plastics industry relies heavily on thermal and thermo-oxidation stabilizers since without their protection during manufacturing, all typical thermoplastics would deteriorate significantly. In plastics, they are usually added in extremely small amounts to prevent the polymer from breaking down too soon during usage and processing. Consequently, not much attention has been paid to their toxicity. The potential toxicity of these stabilizers, however, depends on both their leachability and intrinsic toxicity. Even at relatively low quantities in the plastic, an EDC addition may have substantial toxicity. Solar UV stabilizers should only be applied to polymers intended for outdoor use, as opposed to thermal stabilizers. These are utilized at very low levels, and some of them, like the well-known HALS additives are macromolecular and difficult to leak.

When stabilizer concentrations in plastics are high enough, they can pose risk to human health and the environment. For example, rat and mouse exposure studies have established the carcinogenicity of several benzophenones. Benzophenones are frequently found in transparent plastic items as a light-absorbing or UV stabilizer. Endocrine disruptors (EDCs) have been linked to several photostabilizers in this class, including 2,2',4,4'-tetrahydroxybenzophenone (2-benzophenone) and 2,4-dihydroxybenzophenone (Benzophenon-1) (Andrady and Rajapakse, 2019). Other stabilizers that are found in the environment and are used with thermoplastics but have not been identified as EDCs include amines and benzotriazoles. It is also claimed that the majority of dibutyltin-based stabilizers utilized in polyurethane and PVC compositions are EDCs. Food and beverage packaging accounts for almost half of all packing applications, making the packaging business an important industrial sector. Packaging is crucial to the production of food because it prolongs the shelf life of food items by shielding them from air, gas leaks, moisture loss, light radiation, artificial fragrances, microbiological contamination, temperature fluctuations, and mechanical stresses. Additionally, it makes certain food preparation, storage, and even transportation easier. Foods are packaged using a variety of materials. Nonetheless, plastic makes up the majority of consumer packaging. Diffusion-controlled methods allow food packaging and packaged foods to interact, and different substances may migrate into the food. These compounds are additives that are used during the production process to obtain certain material qualities, as well as initial materials utilized in the first polymerization step, such as monomers or catalysts. Starting materials may leak because of either insufficient polymerization during the material's development or gradual material deterioration. Moreover, these materials or additions may include contaminants, such as degradation products (non-intentionally added substances, or NIAS), which may have again seeped from plastics.

4.7. POLYOLEFINS

The two most used polyolefins are Polyethylene (PE) and Polypropylene (PP). Both are easily processed, recycled, and reused. They are also lightweight, flexible, stable, and resistant to chemicals and moisture. Low-Density PE (LDPE) and High-Density PE (HDPE) are the two varieties of PE products. Milk, juice, and water bottles, and grocery, waste, and retail bags are all made of HDPE. LDPE is primarily utilized in film applications and in applications that need heat sealing, such as flexible lids, squeezable food bottles, frozen food bags, and bread bags. Microwavable and hot-filled packaging items can both employ PP. It's also found in margarine tubs and yogurt containers. PP offers the strength and moisture barrier for ketchup and salad dressing bottles when combined with an oxygen barrier like polyvinylidene chloride or ethylene vinyl alcohol. Polyesters include polycarbonate (PC), polyethylene naphthalate (PEN), and polyethylene terephthalate (PET or PETE). One of the plastics that is most frequently utilized in the packaging sector in Europe is PET. Typical uses for PET consist of soft drink bottles, fruit juice bottles, mineral water bottles; these are particularly good for carbonated drinks, cooking and salad oils, sauces, and detergents; wide-mouth jars and tubs for jams, fruits, and dried foods; trays for precooked meals, pasta dishes, meats, and vegetables that can be reheated in a microwave or conventional oven; foils for boil-in-the-bag pre-cooked meals, snack foods, nuts, sweets, and long-lasting confectionery; and other PET products with an additional oxygen barrier that are perfect for holding beer and vacuum-packed dairy products, such as cheese, processed meats, bag-in-box wines, condiments, coffee, cakes, and syrups. PET is a material that is widely recycled, helping to meet the EU Directive's recycling requirements for plastics. Reusable food and drink containers are among the many typical products that employ PCs. The primary application of polycarbonate is as an alternative to glass in products such large sterilizable infant bottles and returnable/refillable water bottles. PEN is a good material to use when making beer bottles since it prevents the transfer of flavors and odors. After PE and PP, PVC is the most often used plastic.

Figure 4.3. *Food packaging.*

Source: https://en.wikipedia.org/wiki/File:SnackfoodRackDF.JPG jpg, Creative Commons License.

It is utilized in bottles and labels, as well as food and candy packaging material. PVDC, or polyvinylidene chloride, is utilized in multilayer films as a thin barrier layer as well as a monolayer film. Products like chicken, cheese, cured meats, snack foods, tea, coffee, and confections are packaged using it. Egg cartons, containers, casual plastic cutlery, lids, cups, plates, bottles, and food trays are examples of packaging that uses PS for protection. One class of plastics known by the brand name nylon is called polyamides. Because of their hardness and low gas permeability, nylon films are used in food packaging, particularly for boil-in-the-bag food packaging.

4.8. PACKAGING

Food packaging (Figure 4.3) always uses Ethylene Vinyl Alcohol (EVOH) in the form of multilayer structures, with an EVOH film shielded by water-resistant polymers (usually PP) (Mokwena and Tang, 2012). Various additives, including plasticizers, stabilizers, antioxidants, and antistatic and anti-blocking agents, have been developed to enhance the performance of polymeric packaging materials during processing, manufacturing, or use.

However, there is growing concern about food safety, particularly with regard to additives that come from packaging materials into food because it has been claimed that some of them may have harmful impacts on human health. Essentially, rather than based on their chemistry, additives are categorized based on the tasks they carry out. Low molecular weight synthetic organic compounds called plasticizers are added to plastic polymers to enhance their extensibility, flexibility, and processability. Their movement may be impacted by temperature and storage duration. Because of their possible estrogenic and carcinogenic effects, diesters containing phthalic acid have been the subject of numerous surveys and use limitations.

The less hazardous Adipates Acetyltributyl Citrate (ATBC) and Di (2-Ethylhexyl)Adipate (DEHA) are also frequently employed as plasticizers. High concentrations of DEHA have been found in a variety of foods packaged in plasticized PVC films. Although ATBC, which is also used as a plasticizer in flexible PVC films, is not a strong multisite carcinogen. The most common ingredient in plastics is stabilizer.

Epoxidized seed and vegetable oils, such as soybean oil, and organotins are representative examples of stabilizers used in food packaging. Widely utilized as a plasticizer and/or scavenger for hydrogen chloride released from PVC during heat treatment for production, ESBO (Epoxidized Soybean Oil) is a vegetable oil. PVC that contains ESBO is typically used in household food, as well as plastisols for glass jar lid gaskets and organic can coatings. For a long time, organotins have been utilized as polymerization initiators or as additives that stabilize heat and light.

4.9. ORGANOTINS

Organotins are substances that may affect the endocrine system (Graceli et al., 2013). Twenty seven distinct organotin compounds or mixes are permitted for use in plastic-only food contact products under EU regulation. Because specific migration limitations are dependent on tin content, they differ significantly. The most important light stabilizer classes are technically 2-hydroxybenzophenones, 2-hydroxyphenyl benzotriazoles, hindered amines, and organic nickel compounds (polyolefins), along with salicylates, cinnamate derivatives, resorcinol monobenzoates, oxanilides, and p-hydroxybenzoates. Ultraviolet (UV) stabilizers are widely used. The most recent advancement in this area is represented by the sterically hindered amines (HALS), which are frequently used in polyolefins.

In the polymer business, antioxidants are chemicals that are frequently employed to stop polymers from degrading during processing. Polyphenols, organophosphates, and thioesters are compounds that are frequently utilized for that purpose, although their applicability is uncertain because of their toxicity and potential for food migration. Because of the possible hazards, the existence of artificial antioxidants in food is called into question, necessitating stringent legal restrictions. The use of natural antioxidants is suggested as an alternative, including tocopherol, plant extracts, and essential oils from herbs and spices. Fatty acid amides are slip additives that are frequently utilized. High-molecular weight organic compounds like siloxanes are also employed in specific applications, despite amides' widespread use due to their excellent cost structure and high efficiency. It is significant to highlight that the US Food and Drug Administration (FDA) handles both risk assessment and risk management, in contrast to the EU, where these functions are clearly divided between the European Commission (EC) and the European Food and Safety Authority (EFSA).

Figure 4.4. *Codex Alimentarius.*

Source:https://www.fao.org/sustainable-development-goals-helpdesk/champion/article-detail/codex-alimentarius/en, Creative Commons License.

Furthermore, exposure rather than migration is the basis for the regulation of materials that come into contact with food. Along with other pertinent legislation, the FDA implements the Food, Drug and Cosmetic Act of 1958, which is the fundamental regulation on food contact materials. Additives, often known as indirect additives, are compounds that are predicted to contaminate food and hence require clearance.

4.10. CODEX ALIMENTARIUS

Furthermore, the safety, quality, and equity of the global food trade are addressed by the Codex Alimentarius (CA) (Figure 4.4) international food standards, guidelines, and codes of practice (Swinnen et al., 2015). Consequently, the Codex Alimentarius Commission (CAC) recommends sampling plans and maximum values for pollutants and toxins in food and feed, which are listed in Codex Alimentarius. Any substance that is unintentionally introduced to food and is present as a result of food production, processing, preparation, treatment, packaging, transit, as well as environmental contamination, is classified as a contaminant under Codex Alimentarius. In the CAC, food packaging has not gotten much attention. The Codex Alimentarius stipulates that materials and packaging designs must offer sufficient protection for goods in order to reduce contamination, stop damage, and allow for correct labeling. When employed, packaging materials and gases must not be toxic and must not endanger food safety or appropriateness under the given storage and usage conditions. All compounds that fit the Codex definition of a contaminant, including those found in animal feed used for food production, are handled in accordance with Standard 193-1995, which specifies the Maximum Level (ML) of substances that are permissible under law. The IARC has categorized acrylonitrile and vinyl chloride monomers as human carcinogens. Because these pollutants may find their way into food, Guideline Values (GL) of 0.02 mg/kg and 0.01 mg/kg, respectively, have been established for these contaminants. Two categories of migration limits have been established, based on the toxicity, migration behavior, and safety assessment of the approved substances: The maximum amount of a substance that can be released from a material into food or food simulants is known as the specific migration limit (SML, or mg of substance per kg of food). The maximum amount of non-volatile substances that can be released from a material into food simulants is known as the overall migration limit (OML, or mg of substance per kg of food). It is customary to assume that a person weighing 60 kg consumes 1 kg of food containing the chemical per day for determining the SML (Muncke, 2011). Furthermore, it is believed that the food contact surface area is 6 dm^2/kg food and that the 1 kg of food is in touch with a plastic material that releases the substance at the SML. A general specific migration limit of 60 mg/kg is applied to substances for which no particular migration limit or other restrictions are specified. This is classified as SML if the toxicological evaluation yields a specific migration limit of 60 mg/kg.

A result of the toxicological evaluation that is greater than 60 mg/kg is not classified as SML since it exceeds the generic SML.

The total migration limit and a material's inertness are related. According to the regulation, items that come into touch with food cannot release their contents in amounts that could alter the food composition. The threshold has been set at 10 mg of components released per 1 dm2 of plastic food contact material surface area, above which the migration is considered to be a change in the food that is unacceptable. Infants and young children up to three years old are considered a vulnerable consumer group. As such, plastic materials and articles intended for contact with food cannot transfer their constituents to food stimulants in amounts greater than 60 mg of total constituents released per kg of food stimulants, regardless of the size of the packaging. This is a deviation from the previous restriction. The Regulation lays forth specific guidelines for determining whether materials and products that have come into contact with food or those that have not yet come into contact with it comply with migration restrictions. When comparing the migration test results with the appropriate limitations, correction factors might be used. Additionally, special rules apply to multilayer materials and items, which consist of two or more layers joined by adhesives or other techniques.

4.11. CHEMICALS

At various points in the food chain, a wide range of chemicals may find their way into food. Different chemicals in food may come from food packaging and other items that come into touch with food. The intricate chemistry of polymers allows for the incorporation of several substances that may eventually find their way into food. The primary determinants of the risk to human health posed by packaging materials are the degree of this migration and the specific toxicity of the chemicals. The type of food and the type of packaging material determine what materials might impact the food. The degree of direct contact, the length of storage, the pH, the temperature, the exposure circumstances, and microwave radiation, UV radiation, and heat from boiling can all have an impact on the migration. Chemicals employed as monomers, additives (plasticizers, clarifiers, antioxidants) and their breakdown products, solvents, as well as inks, adhesives, and other materials in the final product, are among the materials that might migrate into food. Several of these substances have the potential to be estrogenic agents and, even at comparatively low doses either alone or in combination, can have negative effects, particularly during vulnerable stages of human development. In order to assess migration levels, food stimulants that mimic the chemical characteristics of food are typically used. It is challenging to analyze migrants straight into food because of the complexity of the matrix.

Nevertheless, the approximation using food simulants may overestimate or underestimate actual exposure levels because consumers are exposed to contaminants in foods. The quality of bottled water has been a topic of much discussion lately. Over the past few decades, there has been a steady increase in the consumption of bottled water. According to current standards, bottled water must be analyzed for a few specific physicochemical and microbiological criteria. Nonetheless, concerns exist over the quality of bottled water due to its prolonged storage, non-adherence to prescribed preservation guidelines, and reuse of containers and bottles without proper cleaning or disinfection. Many concerns have been expressed regarding the potential migration of chemicals, intentionally or unintentionally added substances during extended periods of storage, particularly in unfavorable environments. The law states that shields/closures intended to prevent contamination or adulteration must be installed on any container used to package natural mineral waters. As a result, it is very important to monitor the materials used in bottles.

4.12. BOTTLE MATERIALS

PET is distinguished by a restricted assortment of additives and minimal dispersion of possible migrants inside the polymer matrix (Marcinčin, 2002). Other containers, primarily in the form of reusable water carboys, are composed of Polycarbonate (PC), where BPA is the main construction component. Numerous investigations have revealed the existence of organic chemicals in bottled water, which may have an impact on the water's organoleptic qualities and endanger users' health. Odor and taste thresholds may be affected by the presence of organic substances, which may serve as a source of carbon for innate microbes. The issue of xenobiotics in bottled waters is complicated, and there is an ongoing discussion on the provenance of various compounds. Through the production process, from the procurement of raw materials to handling, storing, and distribution, bottled water can get contaminated. This includes the formation of organic compounds under inadequate storage conditions, contamination of the water source in the field, during the bottling process in the plant, and migration of materials from the bottle or the caps into the water. Several investigations have looked into the presence of chemicals, either purposefully or accidentally, in bottled waters and other foods. The most notable instance of plastic components migrating into water is the well-known endocrine disruptor 4,4-dioxydiphenyl-2,2-propane. BPA is a chemical with a substantial production volume. Applications involving BPA can be separated into two groups: a primary group in which BPA is utilized as an additive and a secondary minor group in which BPA is altered or polymerized in its chemical structure. The tough plastic polycarbonate is polymerized with nearly 65% of the BPA. About thirty percent of BPA is polymerized to epoxy resins, which are primarily employed as an interior can coating to shield the cans from rust and corrosion and to prevent direct contact between the food or beverage and the can's metal wall. Moreover, polyester resins, polyacrylates, polyetherimide, and polysulfone polymer all contain BPA. Because of its antioxidant qualities, BPA is added to PVC and thermal paper.

4.13. MORE ON BPA

Even at very low levels of exposure, BPA has been linked to an increased risk of diabetes, cardiovascular disease, malignancies of the reproductive system, issues with fertility, and other endocrine-related problems. The European Commission passed Directive 2011/8/EU in January 2011 limiting the use of BPA in plastic baby feeding bottles (Geens et al., 2012). It is illegal to use BPA-containing polycarbonate baby feeding bottles. A particular migration limit of 600 μg/kg has been set to prevent migration from food packaging materials into food items based on thresholds for probable harmful health consequences of BPA. A safe limit for daily exposure throughout a lifetime is represented by the Tolerated Daily Consumption (TDI) regulated by the European Food Safety Authority or EFSA at 50 μg/kg. Similarly, for long-term oral exposure, the US EPA's Integrated Risk Information System (IRIS) recommended a Reference Dose (RfD) of 0.050 mg/kg bw/day. Because some non-polymerized, free BPA remains after polymerization, BPA exposure may occur through these materials. High temperatures, detergent exposure, and alkaline environments can cause the polymer to break down. Foods and beverages that come into contact with plastics may eventually contain BPA due to the instability of the polycarbonate bond. BPA is readily eluted from polycarbonate products when it is present on or close to the surface. BPA was the only component found in PC polymers, with a mean concentration of 0.748 μg/kg. Polycarbonate plastics, are used in many different applications, including water carboys, reusable milk containers, food storage vessels, and baby bottles. Cans coated with epoxy resin coatings, which are used to coat the inside of food cans, wine storage vats, water containers, and water pipes, are also the main sources of BPA exposure. The kind of food, handling and storage circumstances, and coating/packaging materials all affect the presence of BPA in food products. There is a wide range of reported BPA amounts in canned food. In the Belgian market, canned food had greater BPA amounts than corresponding non-canned food. There was a larger BPA concentration in canned beverages compared to non-canned beverages. It appears that eating canned food is a major source of BPA.

4.14. RESEARCH FINDINGS

Numerous investigations have reported finding BPA in bottled water. BPA levels in PC bottled water varied from 70 to 4,210 ng/L, with freshly made bottles having greater concentrations. Bottle closures, the water itself, or the utilization of recycled PET could be the source of BPA in PET bottled water. The rate at which BPA migrates from polycarbonate containers into water is accelerated when bottled water is exposed to high temperatures and sun radiation. Another form of BPA exposure is bottled water. In the summer, drinking water from polycarbonate containers is associated with consuming bisphenol A. This exposure route, however, only makes up a small percentage of the daily allowance that is tolerable, therefore there is probably little health risk. Concerns were made regarding the potential presence of BPA in tap water due to epoxy pipe linings, which are frequently used to repair drinking water pipes in various nations. Commonly referred to as phthalates (PEs), phthalate diesters are the most important class of plasticizers. PEs are employed to influence the workability, flexibility, and durability of polymers; nevertheless, they are also present in paints, adhesives, inks, and cosmetic products. Approximately 80% of plasticizers produced are utilized in the production of Polyvinyl Chloride (PVC). PVC has up to a 20–50% additive load, which allows for migration to the food supply. DEHP is the most widely used PE, making up around 80% of all plasticizers used in PVC (Erythropel et al., 2014).

In vitro tests have revealed that phthalates including BBP, DBP, DEHP, and DIBP cause estrogenic responses, suggesting that they may have endocrine disrupting properties. Moreover, DEHP may cause cancer in people. The European Community imposed statutory restrictions on plastic and items meant to come into contact with food through the Fourth Amendment to the Plastics Directive in response to the toxicity of numerous PEs and to lessen human exposure. DEHP, DBP, BBP, DiNP, and DiDP are the chemicals of interest. Depending on whether food contact materials are used once or repeatedly, specific migration limits and compositional limitations in plastics have been specified. Plastics that may come into contact with food were restricted by the European Commission. For DBP, BBP, DEHP, and DEHA, the suggested SML values are 0.3 mg/kg, 30 mg/kg, 1.5 mg/kg, and 18 mg/kg, respectively. The maximum levels of DEHP that are allowed in drinking water are 8 μg/L and 6 μg/L, respectively, according to the WHO and EPA. In addition, the EPA recommends closely monitoring drinking water for phthalate at quantities higher than this threshold. Furthermore, other substances like BBP, DBP, DEP, and DINP are included as potential contaminants in the contaminant list under the act named as Safe Drinking Water Act. It should be mentioned that the Maximum Contaminant Level Goal (MCLG) for DEHP in water has been set at zero by the EPA under the Safe Drinking Water Act. The EPA has set this target in order to avert potential health issues,. PEs are released and get migrated into foods and drinking water from the packaging materials because they lack chemical reactivity. PEs as a type of NIAS found in materials that are not contained due to their widespread use. One well-known cause of food contamination is the migration of PEs from items that come into contact with food into foodstuffs. Food is thought to be the most significant source of exposure to DEHP and PEs, making up 80–90% of an adult's daily consumption and sometimes up to 1000 times more than from water.

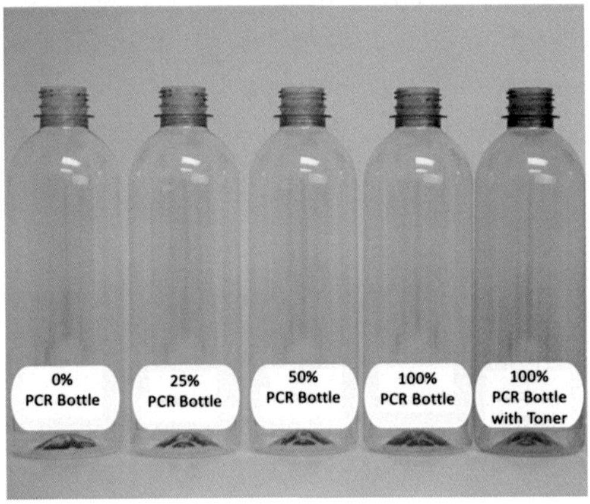

Figure 4.5. *PET bottles.*

Source: Benyathiar, P., Kumar, P., Carpenter, G., Brace, J., & Mishra, D. K. (2022). Polyethylene Terephthalate (PET) Bottle-to-Bottle Recycling for the Beverage Industry: A Review. Polymers, 14(12), Article 12. https://doi.org/10.3390/polym14122366, Creative Commons License.

It is believed that PET bottles (Figure 4.5) are not made using PEs. The contamination that was detected is, therefore, most likely a result of the bottling and production processes, or the cap liner that was employed. An important issue that arises from the widespread usage of plasticizers during analytical procedures is cross-contamination. In general, the levels of PEs found in bottled water were lower than the harmful thresholds suggested for drinking water. Nonetheless, one way to be exposed to PEs is through drinking water. When compared to the allowed daily total intake of PEs, bottled water appears to have a minimal impact (<1.1% TDI) and was deemed acceptable even under the most adverse storage circumstances (Vilarinho et al., 2019). Since youngsters drink more water than they weigh, their exposure to water decreases as they get older. Furthermore, there is little chance of cancer from DEHP in bottled water. Two widely recognized byproducts of PET breakdown are formaldehyde and acetaldehyde. They are produced when PET degrades thermo-mechanically and thermo-oxidatively. Different aldehyde levels were formed in the PET bottle wall for possible reasons related to the manufacturing procedures, molecular weight of the polymer, and chemical quality of the raw material utilized in the production of PET bottles. While acetone, propanal, and butanal are examples of other carbonyl compounds that are included in the list, no specific SMLs are mentioned. Formaldehyde and acetaldehyde were found in PET bottled waters. The kind of container, the temperature and length of storage, the carbonation of the water, and the exposure to sunlight all have an impact on this migration. The migration of these chemicals is accelerated when stored at high temperatures (60°C). This may be explained by the fact that polymeric chains directly related to the migration phenomena become more mobile when stored close to the PET glass transition temperature, which is around 80°C for semi-crystalline PET. The migration of aldehydes in bottle water is also influenced by the type of water. The bottled carbonated waters included a significant amount of formaldehyde and acetaldehyde, most likely as a result of the carbon dioxide pressure on the bottle wall and acid pH. Researchers found no correlation between aldehyde migration and sunlight exposure. Nonetheless, some researchers discovered that carbonated water kept in PET bottles had higher formaldehyde and acetaldehyde contents. Mineral waters that have an off-flavor due to formaldehyde have lost some of their organoleptic qualities. In fact, mineral water packaged in plastic materials developed off-odors similar to those of plastic due to UV radiation exposure.

Food packaging containing styrene is mostly caused by monomers seeping out of polystyrene containers. The production of food contact products/ substances, such as trays for meat, cookies, and sweets along with disposable plates, cups, and other items, uses Polystyrene (PS), which was utilized in almost 50% of food packaging and food service items. Styrene monomers may occur in PS goods during manufacturing and then migrate out of these materials and into food. The lipophilicity of the food, the surface area of the container relative to the volume of food, the length of contact, and the temperature of the food are the key factors influencing the migration of styrene monomers from polystyrene containers. Since it is reasonably expected that styrene will cause cancer in humans, a number of multinational businesses have begun to phase out polystyrene foam packaging from their products.

4.15. ANTIMONY IN BOTTLED WATER

One ingredient that is typically present in bottled water from PET bottles is antimony (Sb). Because antimony trioxide (Sb_2O_3) is effective, has a low propensity to cause adverse effects, and doesn't provide color, it is frequently employed to speed up the production of PET. Sb is usually found in PET at concentrations between 170 and 300 mg/kg. Depending on the production process, there may be differences in the residual concentration of Sb on the PET bottle surface. Water receives a tiny amount of the Sb that is present in PET. Sb has been found in bottled water under a variety of storage circumstances (Keresztes et al., 2009). PET containers showed substantially more Sb leaching than PC, HDPE, and PS containers. The type of water, exposure conditions, temperature and length of storage, and container color all have a big impact on the amount of Sb that leaches from PET into water. The release of Sb is also significantly influenced by temperature. Both short-term and long-term experiments on the migration of Sb from PET bottles into mineral water at varying temperatures revealed an increase in Sb migration at temperatures below 60°C, which is often below the EU limit of 5 μg/L. Sb leaching was also increased when samples were exposed to artificial radiation or sunlight. In carbonated waters, Sb migration was greater; carbon dioxide most likely enhanced this migration. The Sb migration potential and PET color did not correlate. While some researchers found that Sb migration was greater in water held in dark-colored containers than in clear ones, some researchers reported a rise in Sb concentration in clear PET compared to blue bottles. It should be noted that reusing bottles to refill tap water is a typical practice that increases the release of Sb with each reuse cycle. Sb may be found in drinking water. Studies on biomonitoring revealed a correlation between urine Sb concentrations and water use. The way that additives are integrated into plastic materials or polymers can also be used to understand how they are released from them. The polymers and additives are not chemically linked. They are typically mixed into polymers during manufacturing. The polymers and the additional PBDEs are not chemically linked.

The movement of the additive molecule within the polymer as well as its release or migration from the polymer can be controlled by the size of the additive molecule, the internal pore size of the polymer, and the hydrophobicity, or affinity, of the additives to polymers. The capacity of additives to migrate is minimal yet considerable if they fit perfectly into the pores. Nevertheless, a polymer with a larger pore size allows for the easier passage of smaller additives or those with a lower molecular weight. As a result, additives may seep through the surfaces of polymers or plastics. PBDEs leach from polymers differently based on their molecular size and hydrophobicity, among other chemical characteristics. Monomers, like the molecules that make up BPA, are the building blocks of polymers that are made of carbon, hydrogen, and oxygen bound together by covalent bonds. Cleavage of the covalent bond is therefore necessary for the release or migration of a monomer that is chemically or reactively bound from a polymer. Covalent bonds can break via chemical reactions including hydrolysis and oxidation. When exposed to acidic or basic materials over time and at elevated temperatures, the BPA monomer has been reported to be a breakdown product of the BPA-based polymer.

4.16. FLOATING PLASTICS

Even the most isolated regions in the world have been visited by scientists, who are evaluating marine debris in the Arctic. While minimal or negligible levels of litter were observed in Antarctica, marine litter ranged from a few pieces up to 7,719 items/km^2 in the Hausgarten Observatory, with plastics ranging in moderate percentages of 59–65% (Ioakeimidis et al., 2019). It is clear that a substantial marine litter stock, estimated to be between 71.5 and 116 billion big plastic litter items, is present on the seafloors, which exhibit spatial heterogeneity. It is important to notice that the tiny plastic particles that are building up in the deep-sea sediments are not included in these calculations. Thus, it is thought that the deep-sea bottom serves as the last sink for marine debris. Plastic debris in the Mediterranean Sea is primarily building up on the seafloor in the basins to the west and east. This may be related to the fact that the majority of the water masses entering the Mediterranean Sea come from the Atlantic Ocean in the Western Mediterranean Sea, while the currents in the Eastern Mediterranean Sea are weakening and unable to effectively loop back the plastic masses.

There is less information/data available on floating plastic than there is about marine waste on the seafloor. In many parts of the world that have been researched, there is a great degree of spatial abundance variation with regard to size and weight. Seas and oceans are seeing a movement and accumulation of plastic and microplastic particles, primarily due to surface currents that are driven by wind and those that are not. The variability of the Mediterranean surface circulation is primarily responsible for the fluctuating plastic distribution in the Mediterranean Sea. However, researchers clearly showed that the quantity of plastics in the Mediterranean Sea is exactly equal to that of the five subtropical gyres. There is undoubtedly a need for more research on drifting marine debris globally. Of all the seas and oceans, the Mediterranean and Black Sea Basins are the two that have likely received the slightest amount of research. A clearer picture of the amount of plastic litter on the surface combined with data from the seafloor could lead to the development of certain theories about the relative amounts of benthic and floating waste. The methods for removing plastic waste that can be implemented sustainably in terms of expense, effort, and habitat loss should also receive special consideration. For the various types of marine litter items that are found in the sea environment, a standardized global marine litter assessment should have substantial support. All depth ranges, from extremely shallow coastal waters to deep submarine canyons, should be included in the global assessment. All sampling techniques, including commercial and experimental trawlers, ROVs, Manta Nets, and submersible vehicles, should be used. The assessment should be conducted continuously within a framework designed for monitoring.

Chapter **5**

MICROPLASTICS IN FRESHWATER SYSTEMS

Contents

Introduction 100
5.1. Microplastics 100
5.2. Challenges to Studying Particles 101
5.3. Fragments From Degradation 102
5.4. Other Aspects of Polymers 103
5.5. Chemical Interactions 104
5.6. Particles as Pollutants 105
5.7. Risk Management 107
5.8. Detection of Materials and Associated Issues 109
5.9. Surface Interactions 110
5.10. Assessing the Risks 111
5.11. Other Assessments 113
5.12. Ecotoxicity of Nanomaterials 114

INTRODUCTION

Synthetic polymers, commonly known as plastics, which differ significantly from natural Holocene-era materials, have increasingly accumulated in lake sediments due to anthropogenic (human-caused) activities (Dong et al., 2020). As a result, plastics are often considered a marker of the Anthropocene, the current geological age characterized by substantial human impact on Earth's geology and ecosystems. Plastics have long been recognized as major contributors to riverine pollution, and the degradation of these plastics poses potential risks for soil health. Until recently, however, the majority of research on plastic contamination has concentrated on marine environments. This focus has expanded over the past few years to include freshwater and terrestrial ecosystems, as awareness of plastic pollution's broader impact has grown. Numerous studies have noted the limited data available on the prevalence and effects of plastic pollution in freshwater and terrestrial habitats. Microplastics (MPs) refer to small plastic fragments identified through monitoring in various freshwater systems, including riverine beaches, surface waters, and the sediments of rivers, lakes, and reservoirs. These studies reveal that MPs are widely distributed in these environments and occur at concentrations comparable to those found in marine settings, despite less overall data. Ecotoxicological studies have primarily focused on MP consumption by various species and their impact on life-sustaining parameters alongside this monitoring data. Findings suggest that primary MPs, such as polyethylene (PE) and polystyrene (PS) microspheres, may pose a toxic threat to freshwater ecosystems at high concentrations over short-term exposures, although most studies have focused on MPs broadly. Additionally, there is growing concern that prolonged exposure to MPs could lead to bioaccumulation of submicron particles, potentially harming environmental health over time.

5.1. MICROPLASTICS

The word "microplastics" should be reinterpreted to refer solely to particles in the micrometer size range (items <1 mm), while the term "mesoplastic" should be used to refer to items between 1 and 2,500 mm. Microplastics are defined by Lambert et al as being between 1 mm and >0.1 μm (Herrera et al., 2018). Nonetheless, the 5 mm upper limit is widely acknowledged since it might contain a variety of tiny particles that are easily absorbed by living things. MPs can be classified as primary (1°) or secondary (2°) MPs in most cases. Primary MPs are produced specifically for use as resin pellets to create larger things, or directly in cosmetic products as toothpaste and face scrubs. There's no denying that MPs have been around for a while. MPs are regarded as contaminants of rising significance as a result of increased research on the subject since the early 2000s. Plastics will find their way into freshwater habitats from a variety of sources. Land littering is a significant environmental and public problem. It is also a growing source of concern in protected areas where visitor density affects volumes; as a result, actions to lessen and mitigate environmental damage are now required. Furthermore, there may be a greater emphasis on one source in a given geographic area than another due to regional variations in waste management techniques. Similar to bulk plastic products, MPs can infiltrate the environment through a variety of routes, and a route that is significant in one area may not be as significant in another.

Many possible pathways exist for MPs to be released into the environment: MPs passing through Waste Water Treatment Plants (WWTPs) due to their use in personal care products or the release of fibers from textiles during laundry; MPs being applied to agricultural lands; storm water overflow events; incidental release from industrial products or processes; and atmospheric deposition of fibers (Xu et al., 2021). One of the main causes of plastic contamination of agricultural soils is believed to be the use of plastics for crop production, which is regarded as an agricultural emission. The benefits include protecting against unfavorable weather conditions, preserving moisture, which lowers the need for irrigation; limiting weed development and raising soil temperature, which lessens competition for soil nutrients and lowers fertilizer expenditures, enhancing crop yields. But when combined with precipitation events, weathering can make them brittle and difficult to recover, which leads to the material disintegrating. The residues and broken particles can then be washed into the soil, where they can accumulate. There are additional sources, like emissions from factories and building sites. Large amounts of synthetic particles may also be released by automobile tires. With anthropogenic zinc concentrations that are strongly connected to traffic density, these tire particles are acknowledged as a source of zinc pollution to the environment.

5.2. CHALLENGES TO STUDYING PARTICLES

It can be quite difficult to isolate MPs from environmental matrices, especially when working with materials that have a high organic content, like soils and sediments. Likewise, high pigment concentrations and particle and fiber degradation make it difficult to identify synthetic polymers spectroscopically. Because of this, access to advanced equipment is necessary for the detection and analytical validation of MPs. According to recent monitoring studies, MPs are widely distributed in a range of freshwater matrices, much like they are in marine habitats. The Rhine River in Germany has surface water samples with reported MP concentrations ranging from 8,92,777 particles km^2 to 3.9 million particles km^2 at its peak. The Rhine and Main rivers in Germany have sediments with particle counts ranging from 228 to 3,763 and 786 to 1,368 particles kg^{-1} respectively. There have been reports of high surface water concentrations at the Three Gorges Dam in China. These are ascribed to infrastructural problems with recycling and garbage disposal, as well as a deficiency of wastewater treatment facilities in smaller communities. Because their separation and identification are dependent on visual observation techniques, these investigations may understate the real MP concentrations by excluding those in the submicron size ranges.

A large number of plastic items that come into contact with the environment move about. Rather, they will be moved with different durations (resident times) across environmental compartments. For instance, the type of land cover, the distance to a particular river site, and the current weather will all affect the flow from land to river systems. It is simple to notice how plastic litter accumulates at roadside habitats, and in certain nations, frequent roadside grass cutting methods cause discarded objects to be swiftly broken down by mowing machinery. MPs can then be transported from land to water by surface runoff or by dispersing into roadside ditches through cutting action. Because of stranding and other watercourse barriers like groynes and barrages, the movement of bulk plastics and MPs within the riverine system will be limited by its hydrology and the morphology at a specific river site. Since MPs will be disseminated and transported to other environmental compartments more quickly than macroplastics, they may also degrade at distinct rates from larger plastics. For LDPE in a marine environment, the production of MP-associated biofilms has been studied (Gong et al., 2019). Because they prevent exposure to light, transport to sediments and the development of biofilms on MP surfaces may further slow the pace of deterioration. Most of what we now know about the processes involved in plastic deterioration comes from lab research, which frequently focuses on one mechanism, like photo-, thermal-, or bio-degradation. The deterioration of plastics in environmental contexts where many degradation mechanisms coexist is a subject of limited information. When data is available, these investigations have a tendency to concentrate on things like tensile strength changes, molecular structure breakdown, and the discovery of particular microbial strains that work with specific polymer kinds.

5.3. FRAGMENTS FROM DEGRADATION

The definition of the degradation processes takes into account the experimental result as well as the degradation mechanism being studied. Particle production rates, on the other hand, are frequently overlooked. This is significant because polymers that break down into smaller pieces, like PE, typically do not depolymerize easily. Subsequently, these pieces break down into smaller pieces, ultimately transforming into nanoplastics.

It is not an easy task to forecast/predict the rates of plastic fragmentation. The kinetics of polymer degradation have been thoroughly studied in the polymer science, and kinetic fragmentation models have been examined. The distribution of fragment sizes resulting from breakdown events is described by these models. Rate equations that assume that each particle is exposed to an average environment, that mass is the unit used to characterize a particle, and that the size distribution is taken to be spatially uniform can be utilized to explain these processes. Particle form can be explained by averaging all conceivable particle shapes, and these processes can be characterized as linear or nonlinear. These degradation processes are often described by models that are quite complex, but they usually center on chain cleavage in the polymer backbone through the following mechanisms: random chain scission, in which all bonds break with the same probability and are characterized by oxidative reactions; scission at the chain midpoint, which is dominated by mechanical degradation; chain-end scission, which yields monomers and is found in thermal and photodecomposition processes; and in terms of inhomogeneity, where different bonds have different breaking probabilities and are distributed throughout the system. The calculation of degradation half-lives for highly hydrolyzable polymers has also been studied. Further research is necessary to determine whether it is feasible to describe the exponential decline of more chemically resistant plastics.

5.4. OTHER ASPECTS OF POLYMERS

Polymer characteristics including density and crystallinity, environmental exposure levels, and the kind and amount of chemical additions are significant factors that will affect MP degradation and fragmentation (Issac and Kandasubramanian, 2021). The usage of structural elements that are difficult for biodegradation and the intricacy of the polymer structure are two molecular properties that typically prevent degradation. Since the crystalline area is made up of more closely packed and organized polymer chains, crystallinity plays a significant role in this polymer characteristic. Physical characteristics like density and permeability are impacted by crystallinity. This therefore influences their behavior when it comes to swelling and hydration, which impacts the availability of sorption sites for bacteria. While biological substances break down plastic faster, stabilizers like antioxidants and antibacterial compounds work to extend the life of plastics. In general, MP fragmentation into smaller particles, such as nanoplastics, chemical conversion of the plastic fragments, degradation of the plastic fragments into organic non-polymer molecules, and conversion/degradation of these non-polymer molecules into other compounds are the processes involved in environmental degradation. Because the sorption of hydrophobic contaminants to MPs will alter the mobility and bioavailability of such particles, this process is thought to be significant for the environment. MPs are known to concentrate Persistent Organic Pollutants (POPs) like dioxins, PCBs, and DDT in marine habitats. In an attempt to determine whether the metals were indeed linked to the nonremovable fine organic matter found in the pellet samples, fresh polypropylene pellets were suspended in a harbor for eight weeks. It was discovered that the metals accumulated from the surrounding seawater, with Zn showing the lowest concentration at 0.25 $\mu g\ g^{-1}$ and Fe reaching the highest level at 17.98 $\mu g\ g^{-1}$. The current state of knowledge on freshwater and terrestrial ecosystems needs inputs, even though their pollutant composition differs greatly from that of marine environments. Pharmaceuticals, personal care items, flame retardants, and other industrial chemicals are examples of developing contaminants that are likely to co-occur in freshwater environments with MPs. These pollutants enter the environment as components of complicated solid and liquid waste streams.

5.5. CHEMICAL INTERACTIONS

The reversible sorption on polymer matrix surfaces known as physical adsorption occurs without the creation of covalent bonds. Chemical interactions occur between the sorbate and the polymer surface during chemical adsorption (Hüffer and Hofmann, 2016). Depending on how old the polymer surface is, this kind of reaction may create new chemical bonds there. Variations in the localized environment's pH, temperature, and ionic strength can have an impact on these activities. The process of pore-filling, which is reliant on the chemical's molecular size and the pore diameter of a certain polymer structure, happens when hydrophobic contaminants penetrate the polymer matrix. Here, a polymer matrix with bigger pores will allow contaminants with lower molecular weights to pass through easily. The type of polymer, its density and crystallinity, the surrounding environment, and the contaminants it contains will all affect the adsorption kinetics. For example, because the crystalline part of a plastic material is made up of more ordered and densely organized polymer chains, the sorption and diffusion of hydrophobic impurities are more likely to occur in the amorphous material. Polymers having high symmetry, strong interchain hydrogen bonding, and short, repeating unit structures will have a lower sorption potential. Low-Density Polyethylene (LDPE) and high-density polyethylene are two excellent examples. The high branch concentrations seen in LDPE make it difficult for the polymer chains to be readily stacked one on top of the other. HDPE, on the other hand, is the most structurally similar to pure polyethylene and is mainly composed of linear, unbranched molecules. The linearity in HDPE has a higher density and a high degree of crystallinity. When detecting dissolved Polycyclic Aromatic Hydrocarbons (PAH), Polychlorinated Biphenyls (PCB), and other hydrophobic organic substances in aquatic environments, passive sampling systems frequently employ LDPE. In batch sorption studies, the sorption of PAHs to LDPE and HDPE pellets was also investigated. It was shown that LDPE had greater diffusion coefficients than HDPE, indicating that low-density polymers have shorter equilibrium durations. This suggests that PE's strong sorption capability makes it interesting from an environmental perspective. Furthermore, because smaller particles have a larger surface-to-volume ratio and shorter diffusion periods, particle size will affect the sorption characteristics. It will be difficult to isolate and quantify the sorption mechanisms for all of the polymer types currently in use because different polymer classes will exhibit different sorption behaviors based on the kind and concentration of additive compounds the polymer is compounded with and the potential effects these compounds may have on the density and crystallinity of the polymer.

A fraction of pesticides may bind soils irreversibly. An essential first step in examining the sorption processes that exist between various polymer types and co-occurring hydrophobic pollutants is the study of sorption equilibrium isotherms. This will allow the sorption and diffusion connections between MPs and co-occurring pollutants. How much sorbed chemical becomes accessible in the water column as a result of the ongoing disintegration and degradation of MPs or as a result of alterations in the surrounding environment, such as variations in pH, temperature, or system chemistry, is another interesting subject. International surveys and scientific investigations have provided ample evidence of the persistent presence of microplastics in the aquatic environment. It has been noted that more deterioration and fragmentation could lead to the production of nanoplastics, or plastic particles the size of nanoparticles. Microplastics and nanoplastics may have direct ecotoxicological impacts in the environment in addition to acting as vectors for co-contaminants through adsorption (Thiagarajan et al., 2021).

5.6. PARTICLES AS POLLUTANTS

Aquatic organisms may be negatively impacted by the discharge of plastic additives and monomers from the polymer matrix. Even though there is currently little information available about the ecotoxicological effects of nano- and microplastics, their small size raises concerns about the negative effects and dislocation of these particles inside organisms, issues that are frequently discussed in relation to engineered nanomaterials. Comparably, when it comes to the ecotoxicology of nanomaterials, the transportation of co-contaminants and the leaching of soluble chemicals are argued topics. In terms of possible impacts and anticipated behavior in aquatic media, we can compare nano- and microplastics and manufactured nanomaterials. The application of existing test procedures, which were designed for soluble compounds, to the testing of particle suspensions has proven to be one of the major obstacles in the ecotoxicology of nanomaterials. This frequently calls for adjustments to the test, as well as careful consideration of the physical-chemical characterization and data interpretation.

Given that these particles are becoming more prevalent as environmental pollutants, it is important to comprehend how they behave in the environment and how they could affect living things. Testing for ecotoxicity in particles is a paradigm change away from testing for soluble chemicals and necessitates reevaluating current protocols and techniques, including the OECD and ISO standardized approaches. On the one hand, regardless of whether the particles are plastic or artificial nanomaterials, there are often similarities between ecotoxicological testing of particles. Particle sizes of nano- and microplastics are widely varied. Particles with similar sizes are more likely to exhibit similarities in terms of behavior, fate, and impacts. There are some similarities between the possible sources and pathways through which nano- and microplastics and artificial nanomaterials enter the environment. Engineered nanomaterials are precisely constructed and created for certain procedures, or goods. Production can occur through the comminution of bigger materials or by synthesis. This is similar to the process used to make primary nano- and microplastics, such as plastic pellets used as feed in the plastic manufacturing process or microbeads for cosmetic applications. Primary nanoplastics would actually be considered nanomaterials. In Europe, the use of primary microplastic beads in cosmetics was projected to have exceeded 4,000t. However, only a minute portion of the projected total environmental microplastics load is made up of primary microplastics. However, uncontrolled processes including abrasion and degradation of bigger plastic goods and fragments, that is, secondary sources of anthropogenic origin are the principal contributors to nano- and microplastic contamination. These sources include badly managed plastic refuse, which is either dumped into the environment or incorrectly collected and disposed of in landfills, eventually finding its way into the ecosystem by the movement of water or wind. These can be carried to aquatic habitats by wind and surface runoff water. Synthetic fabrics are another significant source, as washing them has been demonstrated to release significant volumes of microplastic fibers into the waste water (Carney Almroth et al., 2018). When comparing artificial nanomaterials to micro- and nanoplastics, the relative significance of secondary sources is distinct because the former are created through regulated industrial processes, while the latter are derived from the bulk material found in the environment. As a result, their release is associated with certain goods or industrial uses, making them similar to primary microplastics.

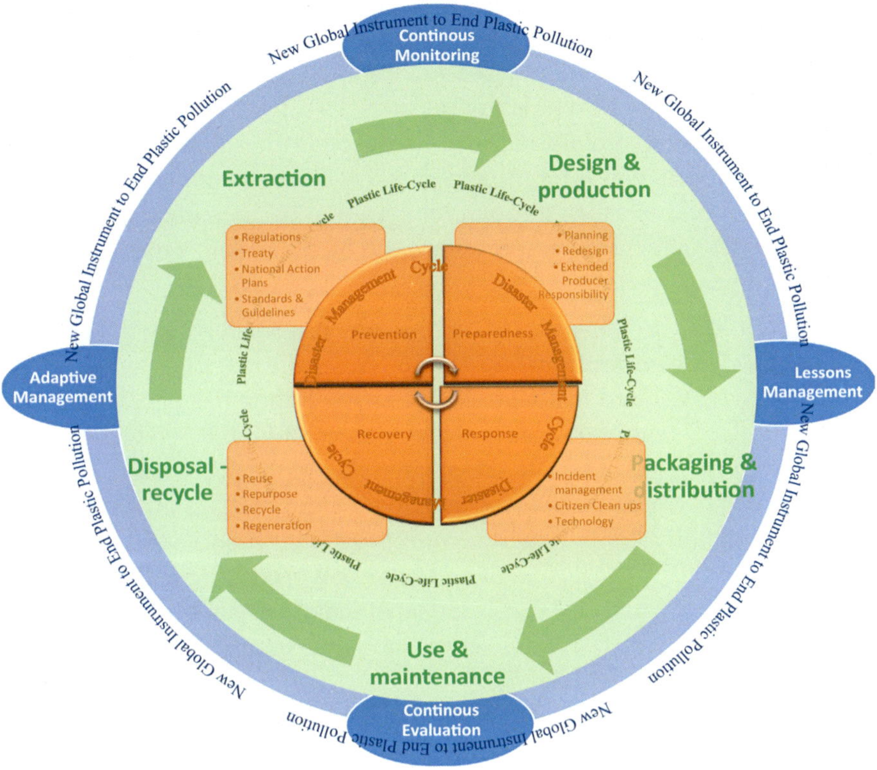

Figure 5.1. *Risk management.*

Source: Senathirajah, K., Bonner, M., Schuyler, Q., & Palanisami, T. (2023). A disaster risk reduction framework for the new global instrument to end plastic pollution. Journal of Hazardous Materials, 449, 131020. https://doi.org/10.1016/j.jhazmat.2023.131020, Creative Commons License.

5.7. RISK MANAGEMENT

Risk management (Figure 5.1) and regulatory choices are affected by the sources of engineered/industrially produced primary particles and accidentally produced secondary particles. By regulating the specific manufacturing and use of engineered nanomaterials upstream, regulatory measures can ensure that risk is minimized to acceptable levels. Reducing the amount of unintentionally created anthropogenic nanomaterials released can be achieved through regulations that control air emissions from different combustion processes. Upstream control may be useful in lowering the environmental emissions of primary microplastics, including micro- and nanoplastics. Examples include the UK's impending ban on microbeads in cosmetics, which forbids plastic microbeads in cosmetics like toothpaste. In contrast, broad measures against plastics entering the environment at all phases of plastic manufacture, use, and waste disposal are necessary to reduce the environmental occurrence of secondary microplastics. One type of regulation meant to lessen the overall plastic load on the environment is the prohibition of single-use plastic shopping bags. Another type of regulation is bottle return programs. It is nearly impossible to reduce the creation of microplastics through regulatory measures once the plastic has reached the environment since it is controlled by the plastic's intrinsic qualities and surroundings.

The chemical composition of man-made nanomaterials and nano- and microplastics clearly differ from one another. Any solid material can theoretically be used to create tailored nanoparticles. Carbon Nanotubes (CNTs) and metal oxides (TiO_2, CeO_2, and Ag) are common sources of higher production volume designed nanomaterials. Organic nanomaterials, on the other hand, can also be produced from polymers, monomers, and lipids. Conversely, synthetic polymers, which are created by the polymerization of different monomers and include a variety of materials like Polyethylene (PE), Polypropylene (PP), Polystyrene (PS), and Polyvinyl Chloride (PVC), are what make up nano- and microplastics. Properties of synthetic polymers vary, including density, porosity, and the amount of non-polymeric additives. Additives can include both organic and inorganic materials and can make up as much as 50% of the mass of polymers overall. Because of this, there are as many differences in nano- and microplastics as there are in combinations and ratios of additives, even though they are made of particular synthetic polymers. These additives have the potential to change a material's characteristics, changing how it interacts with the environment and what consequences it has. Regarding designed nanoparticles, the same holds true. The characteristics of manufactured nanoparticles with a specific chemical composition (TiO_2) vary depending on the crystalline structure and surface coatings used. Simultaneously, a variety of distinct materials and material combinations can be used to create engineered nanomaterials.

Polymers are now exempt from REACH (Registration, Evaluation, Authorization and Restriction of Chemicals) registration, according to European legislation (Williams et al., 2009). Particularly in environmental samples and biota, the features and chemical composition of particles have an impact on the viability of particle detection and quantification. Because manufactured nanomaterials are so small, it is very difficult to find them in the environment. Techniques based on electron microscopy, Mass Spectrometry (MS), and spectroscopy can be used to investigate the behavior of the nanomaterials in the test system in a controlled laboratory setting with known nanomaterials. Even when searching for a known nanomaterial, it is not always easy to apply the same methodologies to the detection and quantification of nanomaterials in a natural environmental matrix. Because of this, there is essentially no monitoring data for manufactured nanoparticles. One of the primary issues is that it can be challenging to extract the particles from the sample in their naturally occurring state because the nanomaterials may be altered during sample preparation.

5.8. DETECTION OF MATERIALS AND ASSOCIATED ISSUES

It is challenging to detect manufactured nanomaterials composed of non-ubiquitous elements (Ag) because of complex sample preparations, matrix interferences, and analytical challenges in differentiating between distinct metal species. Because of their organic origin, nano- and microplastics create complications and restrict the range of analytical choices available when they are present in an organic matrix. Larger-sized fractions are relatively easy to collect or extract, such as density-based sand fractionation or water sample filtration; however, smaller microplastics, particularly nanoplastics, are harder to separate from the surrounding environmental matrix. In addition, secondary nano- and microplastics, which make up the majority of plastic particles in the environment, have irregular shapes because they are the outcome of fragmentation rather than deliberate manufacturing. Furthermore, they frequently have neutral, semi-transparent, or transparent colors. Studies comparing the detection of microplastics in environmental sample samples using Fourier Transform-Infrared Spectroscopy (FT-IR) and stereomicroscopy need to be conducted. The true amounts of microplastics can be underestimated because white and transparent fragments were identified by FT-IR but were difficult to find by microscopy. On the other hand, fibers detected by FT-IR as cotton fibers were misidentified as microplastics by stereomicroscopy, resulting in an overestimation of microplastic fibers with this method.

With the advancement of FT-IR coupled with microspectroscopy, or micro-FT-IR, the spatial resolution was significantly increased and particles as small as a few μm could be identified. The method makes it possible to measure reflectance and transmission. Although the latter can also be applied to thick and opaque particles, the former only produces a higher-quality spectrum on thin samples. However, when employing the reflectance mode, uneven surface features (plastic pieces) may result in refractive errors. Attenuated Total Reflectance (ATR) micro-FT-IR can be applied in this situation to enhance the spectrum's quality. The visual pre-sorting of possible plastic particles used in traditional FT-IR procedures is laborious and error-prone. As a result, it is thought that combining micro-FT-IR with focal plane array detectors may enable high throughput investigation of microplastics in environmental samples. However, at the moment, the method can only be applied to particles bigger than 10–20 μm, and sample preparation requires a lot of work. FT-IR, like many other analytical methods for manufactured nanoparticles, is very helpful in controlled laboratory experiments involving microplastics with known composition. This substance can be found in samples after being added to the spectral library. However, because weathering and chemical changes to the plastics' surface can alter their spectra, it can be challenging to employ FT-IR to detect unknown plastic particles from environmental samples. Another approach that is frequently used to identify plastic particles is Raman spectroscopy. A resolution of less than 1 μm is possible when combined with microscopy, or micro-Raman spectroscopy.

Similar to the advancements being made for artificial nanomaterials, techniques for characterizing and detecting nano- and microplastics in environmental matrices with greater resolution, less time, and high throughput are still being developed. For both types of particles, the optimal analytical methods have similar requirements. These methods should be able to handle complex, heterogeneous samples and should result in little alteration to the physical and chemical state during sample preparation; and provide information on several physicochemical parameters, including chemical composition, size, shape, and other characteristics.

5.9. SURFACE INTERACTIONS

Acting as vectors for other contaminants is one of the potential environmental processes that are frequently mentioned for both artificial nanomaterials and microplastics. Engineered nanomaterials and primary (1°) microplastics will come into contact with other chemicals, including preservatives, surfactants, and active components in pharmaceutical drugs, through their use in consumer goods, medicinal, and industrial applications, among other things. The particles will come into contact with toxins found in waste water streams and landfill leachate through various disposal pathways (Luo et al., 2020). As a result, before, during, and after their intended usage, the particles are intentionally and inadvertently mixed with other chemical compounds. Through this process, a particle that would normally be non-hazardous can carry toxic substances. In addition, when hydrophobic contaminants with low water solubility are sorbed to plastic particles, they become more mobile, potentially impacting their distribution and bioavailability by increasing their transport. In the aquatic environment, it has been demonstrated that manufactured nanoparticles can sorb and transport organic contaminants. Nano and microplastics may also serve as carriers of hydrophobic organic compounds.

Smaller particles will typically have a higher capacity for chemical adsorption due to their higher surface area-to-volume ratio. However, because of their small size, they might be easier for organisms to absorb and possibly even move into other areas. The characteristics of the particle and the pollutant control this vector function. Chemical composition, porosity, size, and surface characteristics are significant particle attributes. Solubility can be increased or decreased by weathering processes. An increased surface area and, hence, a potentially higher sorption capacity are caused by the increased surface roughness. Weathering can also alter the surface charge, enhance hardness, and alter crystallinity in order to counteract this. For example, variations in surface charge brought on by weathering might make certain compounds more soluble and others less soluble. Pollutants may also stick to the surfaces of the particles at the same time. For instance, it has been discovered that nanoplastics may adsorb hydrophobic contaminants; this property may be used to remove chemicals from contaminated water and soil. Thus, the processes of adsorption and absorption may be pertinent to the possible role that nano- and microplastics may play as vectors of pollution. A lot of nanomaterials are made of inorganic materials, or in the case of C_{60} fullerenes and CNTs, inorganic carbon. In these situations, adsorption on the particle surface, as opposed to absorption into the particle matrix, controls the sorption of co-pollutants. Therefore, the amount of accessible adsorption sites on the nanomaterial's surface determines its sorption capacity.

5.10. ASSESSING THE RISKS

Engineered nanomaterials, including nano- and microplastics, have the potential to serve as environmental vectors for co-pollutants. The procedure is contingent upon the particular chemical pollutant, particle characteristics, and the characteristics of the surrounding media, which impact the surface properties and the chemical pollutant's speciation and dissociation. The leaching of substances that were initially a part of the particle matrix is another kind of vector function. This mostly refers to the release of coating materials or metal ions from metal and metal oxide nanoparticles in the case of designed nanomaterials. Polymer additives are also susceptible to leaking from plastic particles. The need to accurately quantify ion release is becoming more and more evident in the field of ecotoxicology of nanomaterials, as the concentration of free metal ions is frequently directly linked to reported biological consequences. Engineered nanomaterials are frequently made with specific functionality, reactivity, or biological effects in mind. As mentioned, accidental anthropogenic processes, as opposed to engineered ones, are frequently the source of nano- and microplastics (Gigault et al., 2021). They are not meant to be biologically active. However, some polymer additions might be used to stop biotic or abiotic degradation. Therefore, while assessing the possible environmental impact of manufactured nanomaterials as well as nano- and microplastics, it is helpful to take into account their intended purpose and qualities. When it comes to non-target organisms, engineered nanomaterials meant for biocidal effects are probably more harmful than those meant for inert properties. Similarly, because these compounds may leak out of the polymer matrix, plastic particles containing flame retardants, plasticizers, or biocidal additives are probably more environmentally dangerous.

Physical interactions between the particle and the organism are being emphasized as a key effect mechanism for both designed nanomaterials and nano- and microplastics. It has been noted that certain kinds of synthetic nanomaterials and nanoplastics stick to the surface of microalgae, possibly producing a physical effect at the cellular level. Among the mechanisms mentioned as possibly significant are blockage of the digestive tract, abrasion of tissues, blockage of invertebrate feeding appendages, blockage of enzyme production, decreased feeding stimulus, nutrient dilution, decreased growth rates, lower levels of steroid hormones, and impaired reproduction. Size-dependent variations in effects are particularly interesting for nanomaterials. The entire goal of nanotechnology is to benefit from the unique qualities that come with smaller sizes. When it comes to designed nanomaterials, this includes, for instance, the unique catalytic properties of some materials at the nanoscale, such as cerium dioxide, titanium dioxide, and gold (Au). These are generally inert as larger-sized materials, but they become reactive when their surface area and particle size increase. Thus, even when a material is generally harmless in its bulk form, there is a tendency for toxicity to grow as particle size decreases. Furthermore, the absorption of man-made nanoparticles into tissues and cells may be facilitated by their small size. In aquatic species, artificial nanomaterials have been observed to cause oxidative stress, tissue damage, reduced growth and development, obstruction of photosynthesis, behavioral abnormalities, and increased mortality.

Smaller polymer particles would have unique characteristics due to their higher surface-to-volume ratio. A higher percentage of the molecules will be on the surface of the particle as the particle size decreases. Chemical and biological interactions may increase as a result, since the surface is the site of interactions with the surrounding environment. There are also many absorption pathways by which they enter organisms depending on the size of the particles. To determine the level of risk associated with nano- and microplastics, laboratory-based studies that examine the impacts of these plastics under carefully controlled circumstances must be conducted. Many controlled laboratory studies have been conducted to examine the effects of nano- and microplastics on freshwater organisms. The effects have been detected at several levels, ranging from molecular and cellular to physiological. These consist of higher mortality, behavioral abnormalities, lipid and amino acid metabolic disturbance, poor growth and feeding rates, inflammation, and impaired reproduction. There is a wider range of biological consequences of nano- and microplastics that have been discovered when studies involving marine organisms are included. The majority of studies include diverse characteristics, such as test species, exposure time, exposure concentration, and response variables, and particle type, that is, different polymers, sizes, shapes, and chemicals. This hinders repeatability and makes it challenging to compare study outcomes. As a result, using standardized tests can be helpful. These tests have several advantages, including controlled and repeatable test designs and inter-laboratory comparability. Standardized ecotoxicity tests also benefit from the vast knowledge that has been accumulated over decades of evaluating the effects of chemicals on certain model organisms. This has been mentioned as a reason for the ecotoxicology of nanomaterials to use standardized short-term experiments as a starting point to understand the fate and bioavailability of man-made nanomaterials in the environment. Other test parameters can be changed separately and under control with the use of a well-defined test system and a completely specified synthetic medium, giving insight into certain processes and mechanisms.

5.11. OTHER ASSESSMENTS

Particle behavior and characteristics can alter over time or in response to interactions with test organisms and released biomolecules. Consequently, it has been acknowledged that performing a particle and exposure characterization both before and during a laboratory test is crucial when working with designed nanomaterials. This provides information on exposure in both qualitative and quantitative terms by analyzing the surface area, size, and surface chemistry of the tested particle as well as the behavior of aggregation/agglomeration, sedimentation, and dissolution in the test system. Furthermore, the rationale behind thorough characterization and carefully considered sample preparation methods relates to data interpretation and avoidance of the introduction of test artifacts. The leaching of molecules from particles is another factor that must be considered. The OECD is now working on test criteria for engineered nanomaterials to examine the dissolution of metal ions from metal-containing nanoparticles. The test findings for plastic particles may be impacted by adherent co-pollutants and compounds that seep from the polymer matrix, such as additives or monomers. The actual release of plastic additives from nano- and microplastics under pertinent conditions must therefore be investigated using appropriate test methods, and as a reference, a control for the effects of chemicals and released additives or adhered pollutants must be included.

The integrity, behavior, and persistence of nanomaterials in aquatic media can be strongly influenced by a variety of transformation processes, including oxidation/reduction, interaction with macromolecules, light exposure, and biological transformation. The degree of degradation can be increased or decreased based on the particular conditions. For instance, enhanced dissolution could make metal and metal oxide nanoparticles more hazardous (Odzak et al., 2014). Particle size may also gradually decrease as a result of it. Aging and weathering processes for nano- and microplastics should also be taken into consideration since they can alter the attributes of the particles and promote fragmentation. It is important to note that no evidence of a full breakdown of plastic particles in environmental settings has been shown. Although the aging processes and kinetics associated with nanomaterials, as well as nano- and microplastics, may vary, aging is potentially significant for these materials in the environment. From what is now known, some designed nanomaterials (metal and metal oxide nanomaterials like ZnO and Ag) may release more soluble chemicals and have a lower core persistence than nano- and microplastics. Future studies in this field are obviously needed, both for nanomaterials and for nano- and microplastics.

5.12. ECOTOXICITY OF NANOMATERIALS

The field of ecotoxicity testing of nanomaterials has benefited from the availability of typical industrial nanomaterials and reference materials for the development of test methods. These materials are useful for undertaking similar inter-laboratory and inter-species experiments, as well as validating analytical methods. Similar benefits would accrue to the field of ecotoxicity testing of nano- and microplastics from the establishment of sources of well-characterized, ecologically and industrially relevant materials of different sizes and compositions. There have been concerns raised over the suitability of the current standard ecotoxicity testing for man-made nanomaterials. Concerns with evaluating nano- and microplastics are the same. Their dynamic character in environmental media poses a unique issue as it might lead to variations in relevant exposure routes and potentially distinct impact mechanisms. By diffusion, soluble compounds can enter aquatic organisms and subsequently be transported throughout them to lipid tissues. The process of soluble chemical uptake by cells often involves the crossing of biological membranes, either by passive diffusion or active uptake methods like carrier-mediated transport or ion channel transport. They can operate either non-specifically in the tissues, causing narcosis, or particularly in the tissues, preventing or altering specific biological processes. Rather than molecular diffusion, mechanisms like feeding are responsible for the uptake of particles by organisms. Particles may be absorbed by cells through mechanisms like phagocytosis. Because of this, effects will probably be different from those of soluble substances.

The goal of ecotoxicity testing is to create a dose-response relationship based on the fundamental premise that exposure time, dose, and concentration have a significant impact on outcomes. However, for manufactured nanomaterials, the connection between agglomeration size and concentration has been found to be inverse, indicating that larger agglomerates tend to develop at larger particle concentrations. Excessive levels of designed nanomaterials have also been connected to effects that are brought on by overdosing the test organisms with engineered nanomaterials, which results in physical inhibition rather than a toxic response. Multiple cellular changes must occur before an organism exhibits impairments to its survival or reproduction. Thus, compared to responses of the entire organism, cellular responses may be more vulnerable to microplastic particle stress. It is difficult to identify and define a significant collection of response variables on this level, though, because we are working with a complicated network and a large number of reactions. In order to utilize cellular responses as markers of the potentially dangerous characteristics of nano- and microplastics, more study is required to create appropriate testing procedures. Chronic effect studies are another method for determining the toxicity of relatively low particle concentrations since they can yield more sensitive results than acute ecotoxicity.

Plastics are susceptible to weathering processes once they are in the environment, and these processes will affect the microplastics' surface. The surface textures of microplastics can be exploited to identify the oxidative and mechanical weathering processes (Sun et al., 2020). SEM is typically used to study surface textures. While granules are thought to be the product of oxidative weathering, characteristics like grooves, and cracks are thought to be the product of mechanical weathering. FTIR can be used to confirm the surface oxidation of polymers since it shows the occurrence of peaks for carbonyl groups. When researchers looked at the surface textures of microplastics from the shoreline sediments of the Siling Co basin, they found that mechanical weathering characteristics were more frequently seen. This was explained by the windy weather. This result deviates from that of microplastics recovered from Hawaiian beach sands, but it is consistent with the general trend of microplastics recovered from the Great Lakes. Hawaii may encourage the oxidative weathering of the polymers because it has a warmer and more humid climate than northern Tibet and the Great Lakes region. The making of plastics involves the use of several different polymers. The polymer types that plastic materials are composed of have a significant impact on their performances and properties. Thus, the buoyancy of microplastics can be greatly influenced by polymer types, which in turn can alter how they behave in the environment.

The rate of growth has unavoidably put pressure on urban services, not the least of which is the management of solid waste, where municipalities in developing nations frequently lack funding, have subpar waste collection equipment, and lack urban planning. Plastics are included in this refuse. As is well known, plastics are utilized in a wide range of goods, such as bottles, bags, packaging, and numerous more transient items that are thrown away after a year of manufacture. Plastic recycling processes are well-established in certain regions of the world, but even in African nations where reuse and/or recycling programs exist, they are frequently carried out haphazardly due to a lack of legal support. Therefore, a large amount of solid waste is illegally dumped or ends up in landfills. Plastic waste therefore has the potential to infiltrate the aquatic environment when it is close to freshwater systems, where further degradation can generate MPs. Activities related to freshwaters, such as fishing and tourism, may also contribute to MP concentrations. The possibility of MP pollution in these waters is high given the size of Africa's freshwater bodies and the populations and activities they sustain. Conversely, studies have been done on the existence of MPs in the marine and estuarine environment, and several nations have made great strides toward eliminating the use of plastic bags. Not only has this success been made in reducing plastic waste, but it has also improved public health and the environment. The improper disposal of plastic bags has been demonstrated to obstruct gutters and drains, causing storm water issues. Additionally, the accumulation of water serves as a dock for mosquitoes that transmit malaria, and the use of bags as toilets has been connected to the transmission of diseases. The usage of plastic bags was subject to tax enacted by the South African government in 2003, Rwanda banned the use and importation of plastic bags less than 100 microns in thickness in 2005, and Tanzania followed in 2006 (Nielsen et al., 2019). Such actions might not always be effective because, in South Africa, taxes were not expected to have a negative impact on the stream of plastic bag litter. As a result, even while the efforts are positive, they might not have much of an effect on the possibility of MP pollution in freshwaters in Africa. But first, the problem's scope must be determined, and research done in marine and estuarine environments may provide the answer in this regard. Substantial information has been acquired regarding the existence, origins, and fate of plastics and MPs in the coastal regions surrounding South Africa and their biota, in contrast to the rest of

Africa. Fragments, threads, and foamed plastic particles were often observed forms, with polyethylene being the most common polymer. A reexamination that was carried out at fifty beaches in South Africa between 1984 and 1989 discovered a noteworthy rise in the average MP density, which increased five years later. The distribution of MPs was analyzed, and it was discovered that the differences in abundances between beaches were caused by inshore currents as opposed to local sources. In contrast, local sources influenced macroplastics more than any other source.

Chapter 6

TOXIC EFFECTS OF PLASTICS

Contents

Introduction..120
6.1. Health Problems Due to BPA..121
6.2. BPA in Consumer Products-What Does Research Say?....................121
6.3. Observations and Effects on Cancer...123
6.4. BPA – Does It Adversely Affect Human Fertility?.............................124
6.5. Endocrine Interference of BPA..125
6.6. Phthalates-Threat to Health...127
6.7. Other Toxic Materials..130
6.8. Consumption Patterns of These Contaminants.................................131
6.9. Presence in Household Articles...132
6.10. Engineered Nanomaterials...135
6.11. Other Synthetic Polymers...136

INTRODUCTION

Several consumer products contain significant levels of the endocrine disrupting chemical Bisphenol A (BPA) (den Braver-Sewradj et al., 2020). Since 1957, BPA has been employed in the commercial production of durable polycarbonate polymers and epoxy resins, which are used in dental resins, and food container linings. Producers generate around 8 billion pounds of BPA annually, making it one of the most widely produced industrial compounds globally. Numerous consumer goods, including toys, laptops, household appliances, food and beverage containers, eyeglasses, and medical equipment, are manufactured of plastics that contain BPA.

Figure 6.1. *BPA-its use in water storage.*

Source: https://en.wikipedia.org/wiki/Bisphenol_A#/media/File:Polycarbonate_water_bottle.JPG, Creative Commons License.

Dental work and metal coatings for food cans, pipes, automobiles, dairy equipment, office equipment, and other metal items are two applications for epoxy resins that contain this chemical. BPA and its derivatives are also utilized in flame retardants (tetrabromobisphenol A), and technical applications like laminates for printed circuit boards. Plastics bearing the recycle codes "3" or "7" may contain BPA. Air, soil, water, landfill leachate, and human tissue have all been shown to contain BPA. It has been demonstrated that the chemical seeps into food and drink items from certain food packaging materials such as polycarbonate containers and the epoxy lining used in metal cans as well as reusable containers. BPA can also be inhaled, come into contact with the skin, be present in dental fillings, and we can be exposed to it at work. Human serum, milk, saliva, urine, and amniotic fluid have all been reported to contain BPA (Figure 6.1).

6.1. HEALTH PROBLEMS DUE TO BPA

Numerous health issues, such as infertility, weight gain, behavioral abnormalities, early puberty, prostate and mammary gland malignancies, cardiovascular impacts, obesity, and diabetes, have been linked to BPA in both human and animal studies. Because BPA could connect to steroid receptors, it was thought to have estrogenic properties. In the 1930s, while researchers were looking for synthetic estrogens, these characteristics of BPA were initially shown in experiments using ovariectomized rats. But when it was discovered that Diethylstilbestrol (DES) was a far more powerful estrogen, BPA was dropped for use in pharmaceuticals. Since then, biochemical tests have demonstrated that BPA attaches to both Estrogen receptor alpha (Era) and Estrogen Receptor beta (ERb), with ERb binding affinity being about ten times higher. BPA does fit within the ER binding pocket. However, its binding affinity for both ER isoforms is almost 10,000 times less than estradiol. BPA can act as an antagonist at lower concentrations and act as an agonist at higher concentrations. BPA also binds to membrane-bound forms of the ER (mER) and with high affinity to a transmembrane ER receptor called G protein coupled receptor 30 (GPR30). These paradoxical actions are probably cell-type specific and may be partially mediated by BPA activation of the AF-2 domain of ERa, not the ligand-binding domain. There is growing evidence that BPA interacts with other nuclear receptors in addition to its estrogenic activity, albeit at higher dosages. BPA binds to the thyroid hormone receptor or TR less strongly than the ER. Additionally, research has demonstrated that BPA binds to the ubiquitous aryl hydrocarbon receptor (AhR), and studies have reported that AhR promotes toxicity through many signaling pathways. Research indicates that BPA and its derivatives induce adipocyte development and adipogenic marker genes in preadipocytes via a variety of methods, hence acting as obesogens.

6.2. BPA IN CONSUMER PRODUCTS-WHAT DOES RESEARCH SAY?

Determining whether consumer products contain BPA and how much of it is leaked into food and beverages under typical settings has been the focus of numerous studies since the 1990s. Studies have specifically looked at the amounts of BPA released from baby bottles, food-contact materials, and epoxy resins, which are used to line metal food cans (Geens et al., 2011). These investigations verify that significant amounts of BPA are present in most canned goods. Scientists believe that current human exposures to canned and bottled goods could be in the low microgram per kilogram range for bottle-fed infants and the nanogram per kilogram range for children and adults due to the wide range of amounts that were recorded. Nonetheless, the hazards of exposure to newborns have decreased due to recent actions taken by manufacturers in the United States (U.S.) and Europe to eliminate BPA from baby bottles and infant feeding cups. Recent research has demonstrated that dental operations may increase human exposures. Determining the health impacts of BPA depends on being able to detect exposure to the chemical with accuracy. Because it incorporates exposure over a recent period, measuring BPA in urine is generally regarded as the most reliable indicator of BPA exposure. In contrast, BPA concentrations in blood are thought to reflect only current exposures because of the chemical's short half-life and evidence that BPA does not bioaccumulate. Serum tests, on the other hand, are currently the most useful method for determining the precise amounts of unconjugated BPA, or free BPA, at any given time. Free BPA is the form that is thought to be more physiologically active since it can attach to nuclear receptors such as ER. Further research is necessary in this area because some argue that the fast metabolism of BPA in the liver and gastrointestinal tract during metabolism leads to very low levels of circulating parent BPA, which is biologically active, entering the bloodstream, preventing BPA from causing disease.

These findings have sparked discussion regarding the possibility that contamination is compromising free BPA blood measurements in the scientific community, even if consistent results have occasionally been reported across investigations. Samples from syringes, containers, tubing, and even water used in research may contain trace quantities of BPA. Furthermore, it's possible that the tools used to assess BPA levels are not calibrated correctly. Today, a multi-agency initiative headed by the National Institute of Environmental Health Sciences or NIEHS is working to enhance measuring techniques for blood BPA detection. Because BPA has a short half-life, hourly variability of BPA urinary concentrations is expected, depending on the timing of urine collection and the kind of BPA exposure. Given that the primary route of BPA exposure is thought to be diet, which most adults alter daily, data about the temporal variability of urinary BPA concentrations determined from a 24 hour urine collection is relevant. For instance, a recent investigation from the U.S. Centers for Disease Control and Prevention or CDC revealed that spot urine samples taken from the same individual exhibited significant variation in BPA amounts between days. This variation emphasizes how crucial it is to choose the right time of day to collect spot urine samples when designing studies to assess BPA exposure by biomonitoring. Better exposure assessment techniques, such as gathering and pooling multiple spot urine samples, implementing 24 hour urine sampling, or developing new matrices for measuring non-persistent chemical exposure, could overall improve epidemiological and biomonitoring studies.

6.3. OBSERVATIONS AND EFFECTS ON CANCER

Some authors contend that contamination is most likely the cause of the detectable free blood BPA levels found in numerous human studies. However, some argue that the presence of free BPA in the blood may also be explained by other exposure pathways, such as sublingual absorption and cutaneous contact. Any matrix can produce trace sample contamination, but since urine has greater detectable amounts of conjugated BPA than blood, the effect is probably less severe in urine. In fact, according to recent pharmacokinetic research at the NIEHS Clinical Research Unit, urine has total BPA concentrations that are 40–100 times greater than blood (vom Saal and Welshons, 2014). Overall, the data clearly shows that the best way to quantify daily total exposure to BPA is to employ 24 hour urine collection. Many strong hormonal messengers that function on receptors that have been conserved throughout evolution are necessary for normal human growth, development, and equilibrium. Exposure to endocrine disrupting substances like BPA may interact with hormonally mediated pathways such as the ER, TR, and AR, increasing the risk of disease. Numerous animal experiments conducted in the past few years have shown links between low-level BPA exposures, that is, exposures below the legal no-observed-adverse-effect threshold, or NOAEL and unfavorable health outcomes. These results lend credence to the idea that exposure to BPA may be linked to some unfavorable health consequences in people. An increasing amount of research on humans has shown links between BPA exposure and disease. Given the specific susceptibility of a fetus and infants to environmental toxicant exposures, exposure during the early stages of life is especially concerning; nevertheless, persistent low-level exposures may also raise the risk of some adult health consequences.

Due to the high level of BPA exposure in humans, even minor side effects could have a significant impact on public health. Research assessing BPA's role in human carcinogenesis has been prompted by the pervasiveness of BPA in the environment and the rise in endocrine-related cancers, including breast, uterine, and prostate cancers. There was no discernible link between higher BPA exposure and the prevalence of general cancer. Many studies have looked into the possible effects of BPA specifically on human breast cancer and mammary cancer in animal models because the most frequent kind of breast cancer is estrogen-sensitive (ER-positive tumors). It was not discovered that blood BPA levels were related to an augmented risk of breast cancer. Many studies conducted on mice have shown that early-life exposure to BPA can disrupt development, which raises the risk of cancer in later life. For instance, a number of studies have demonstrated the effects of maternal administration of low doses of BPA on the development of the prostate in male mouse fetuses, as well as a permanent increase in prostate size and number of ARs in adulthood. More recent research has shown that when males are treated with estrogen in adulthood to replicate the increase in estrogen that occurs with aging in men, treatment of neonatal rats with modest doses of BPA orally or by subcutaneous injection resulted in early-stage prostate cancer. A large body of research demonstrates that BPA exposure modifies the development of the mammary glands in rodents, impacts their function as adults, and raises their chance of developing neoplastic tumors. For instance, research on mice and rats indicates that developmental exposure to ecologically relevant BPA levels during breastfeeding and gestation may alter the development of the mammary glands, potentially resulting in neoplasms. BPA exposure causes adenomyosis, leiomyomas, atypical hyperplasia, and cystic ovaries. Additional study is required to close the information gaps separating its carcinogenic effects in animals from its connections with human cancers. Prospective epidemiological studies that take into consideration the prolonged latency between chemical exposure and the development of cancer are necessary for this.

6.4. BPA – DOES IT ADVERSELY AFFECT HUMAN FERTILITY?

Much research has looked at the connections between male BPA exposure and reproductive outcomes. Semen quality and sexual function are among the study's outcomes. Urinary BPA concentrations and serum reproductive hormones have been linked in some epidemiological studies, however, the direction of the interaction between BPA and male hormones has varied. Increases in urine BPA concentrations have been consistently linked with decreased semen quality. Some studies discovered a correlation between urine BPA concentrations and occupational BPA exposure and lower sexual desire, erectile function, and difficulty ejaculating in Chinese men (Liu et al., 2015). These results imply that male sexual dysfunction and decreased semen quality may be linked to recent BPA exposure. However, the clinical significance of these findings is yet unknown because of the cross-sectional designs of the studies and the inadequate evaluation of possible other exposures. Studies on humans have also revealed links between BPA exposure and anomalies related to female reproduction, such as impacts on the ovary, uterus, and quality of oocytes. There are few studies examining how BPA affects the ovarian shape, and most of them concentrate on Polycystic Ovarian Syndrome (PCOS). Serum BPA levels were found to positively correlate with PCOS. BPA exposure was linked to a drop in peak blood estradiol levels before oocyte retrieval in women undergoing In-vitro Fertilization (IVF), which may have a detrimental impact on the quality of the embryos. Serum or urine BPA levels have been compared to egg quality in a few human investigations. There was no correlation between serum BPA concentrations and oocyte maturation. Raising urine BPA content was linked to fewer retrieved oocytes in women undergoing IVF. These findings imply that BPA may harm oocyte maturation and fertilization, which could lead to a reduction in egg yield during IVF and a poor outcome for the patient. The effects of BPA on blastocyst and embryo development have not been extensively studied. Furthermore, increasing urinary BPA concentrations in the male partner, but not the female, decreased the odds of a high embryo cell count and embryo fragmentation score, suggesting the embryos were of low quality for IVF treatments. Blood BPA levels may be linked to the development of endometriosis. However, the study was limited by a small sample size and the use of methods for measuring serum BPA. BPA exposure should be assessed in future endometriosis risk research utilizing 24 hour urine samples. Higher quartiles of urine BPA concentrations were linked to higher risks of implantation failure (Shen et al., 2020). It is well-recognized that early puberty increases the chance of developing breast cancer in later life. Researchers have begun to investigate whether early BPA exposure in animal models can affect the onset of puberty, mammary development, and subsequent risk of mammary neoplasia, as well as whether higher levels of BPA exposure can increase the incidence of advanced puberty in young girls. There was no correlation found between higher quantities of BPA in the urine and the development of pubic or breast hair. Body mass index (BMI) changed the relationship between BPA and breast development, with females with lower BMI having a stronger correlation between delayed breast growth and higher urine BPA contents.

6.5. ENDOCRINE INTERFERENCE OF BPA

Numerous studies conducted on laboratory animals have indicated that exposure to BPA can interfere with various metabolic processes. This implies that BPA exposure may lead to an increase in body mass at environmentally relevant doses, which could exacerbate the obesity epidemic in humans. The research groups concluded that while the available human data on BPA and diabetes were insufficient to draw meaningful conclusions, animal and in vitro studies propose that BPA may affect glucose homeostasis, insulin release, cellular signaling in pancreatic b cells, and adipogenesis. Urinary BPA content has been linked to an increased risk of obesity in people. BPA exposure, as determined by urine BPA levels, is linked to overall obesity in the adult population in the United States. There is a correlation between children's weight gain and concurrent BPA exposure. Additional research has shown links between adult diabetes, cardiovascular disease diagnoses, abnormalities in liver function and urine BPA concentrations. The cross-sectional design of all these results limits their interpretation. Due to the lengthy latency periods associated with metabolic disorders and Cardiovascular Disease (CVD), concomitant urine BPA concentrations may not be a reliable indicator of BPA exposure during the key etiologic windows for the development of these conditions. Given that obese people have a higher risk of CVD and other metabolic diseases, as well as a higher likelihood of consuming packaged and processed foods containing BPA, it has been proposed that additional factors could be at play and explain the observed relationships. Long-term human epidemiological studies with stronger biomonitoring, such as many 24 hour urine samples, are necessary to ascertain how BPA exposure affects human metabolism.

The neuroendocrine circuitry in the brain that regulates sex-specific physiology and behavior in vertebrates is heavily influenced by steroid hormones (McCarthy and Ball, 2008). As a result, endocrine disrupting factors may cause abnormal behavior that is irreversible. Kids between the ages of 8–11 who now have higher urine BPA concentrations also have higher teacher-reported learning difficulties ratings. The relationship between measurements of BPA exposure during pregnancy or childhood and anxiety or depression has been the subject of several investigations. Children who were exposed to more BPA as youngsters had higher levels of anxiety. While reporting other neurobehavioral effects, researchers did not find a correlation between urine BPA concentrations in children and anxious/depressive behaviors, despite identical levels of exposure. Exposure to bis-GMA-containing dental composites was linked to higher self-reported anxiety and depressed behaviors. Concurrent child BPA concentrations were also shown to be strongly correlated with parent-reported anxiety and depression problem ratings in children between the ages of 8 and 11 years old. There is growing evidence that certain neurobehavioral diseases, including Attention-Deficit/Hyperactivity Disorder (ADHD) and Autism Spectrum Disorders (ASDs), may be influenced by environmental toxins, such as BPA. Some researchers found no correlations between amalgams containing bis-GMA and issues related to ADHD. Some scientists found no correlation between urine BPA concentrations during the first three years of life and parent-reported ADHD behaviors. At three years old, a child's urine BPA concentrations were linked to parent-reported elevated emotional reactivity ratings in children; however, this association was not gender-specific. Researchers found a favorable correlation between parent-reported attention issues and concurrent child BPA urine concentrations.

Children's hyperactivity and aggression were strongly correlated with gestational BPA concentrations; however, this correlation was larger in girls than in boys. The size of the correlation was comparable to that of other environmental toxicants and neurodevelopment, such as IQ/behavior. A class of substances known as phthalates is frequently found in plastics and personal hygiene items. Despite being referred to as phthalates, there are numerous varieties of phthalates in common commercial use, each having unique physical and chemical properties, major applications, and toxicologic profiles.

6.6. PHTHALATES-THREAT TO HEALTH

DEHP and Butylbenzyl Phthalate (BBzP) are frequently utilized as plasticizers in flexible vinyl materials, which are found in a diverse product, including medical devices, food containers, and vinyl flooring. Lower molecular weight phthalates, such as diethyl and Dibutyl Phthalate (DEP), are employed as solvents in paints, glues, personal hygiene products, and time-released medications. Because phthalates have a weak bond with the substrate, they can be readily leaked from items and released into the environment. Some of the most prevalent chemicals are found in indoor air and home dust, these high chemicals are found in food packaging, foods, and a wide range of consumer products. It is not unexpected that over 98% of Americans are thought to have been exposed to certain commonly used phthalates, even though urine quantities of various phthalate metabolites usually vary. Variations in the sources and quantities, individual behaviors, and the short biological half-lives of the chemicals are the main causes of the diversity in phthalate concentrations in biological samples.

Certain phthalates are thought to be endocrine disruptors, which means they interfere with natural hormonal communication and operation. DEHP, DBP, BBzP, and DINP are anti-androgenic substances that negatively impact the developing male reproductive system by inhibiting testosterone and insl-3 (Insulin-like 3) production during fetal development (You and Song, 2021). Studies on animals have also revealed detrimental impacts on both male and female fertility. There is also some evidence of impacts in humans on metabolism, neurological development, asthma, and allergies. Although DEP is frequently used in personal care products, especially as a fragrance ingredient solvent, it hasn't been proven to have any anti-androgenic endocrine disrupting properties. The most used method for measuring human exposure to phthalates is urine analysis performed chemically to look for oxidative metabolites and monoesters. Because phthalates are present in many parts of the environment, sample collecting procedures, and laboratory equipment, accurate readings must be taken with extreme caution. Urine monoesters and oxidative metabolites are the most dependable exposure biomarkers. Equipment used in laboratories or for sampling may contaminate measurements of the parent substance. Furthermore, these pollutants can be changed into monoesters following sample collection by hydrolytic enzymes present in blood, breast milk, and other biological matrices like amniotic fluid and meconium, but not in urine. Since oxidative metabolites do not arise as a result of sample contamination, measurements of these compounds in blood are regarded as reliable. Since phthalates are rapidly eliminated within hours, mostly in the urine, urinary levels of metabolites related to phthalates are thought to be the indicator of exposure. Measuring the amount of phthalates in urine integrates exposure by eating, inhalation, and skin absorption, making it a useful biomarker for epidemiological research of health impacts. It appears that different phthalates contribute differently to different routes of exposure; for example, some phthalates are mostly found in food, whereas others are found in personal care products. Urine levels vary during the day and since phthalates have short half-lives, therefore the best way to determine average exposure levels is to utilize 24 hour urine samples, preferably from several days.

For a wide range of demographics, regions, and age groups, urinary concentrations of phthalate metabolites are available, and these offer helpful details on the relative exposure levels within these categories. Urine levels have been utilized as a reference for back calculating exposure intake levels using pharmacokinetic models and other methods. Utilizing these models, scientists calculated that the average daily consumption of DEHP in the general population of the United States in 2001 was between 0.6 and 2.2 micrograms/kg (Rodgers et al., 2014). Based on these correlations and a cumulative evaluation of exposure to three prevalent phthalates (DEHP, DnBP, and DiBP), other researchers found that 24% of German children had intakes above the Tolerable Daily Intake (TDI), a health-based recommendation. It is impossible to determine which phthalates are likely to be present in ordinary products because they are not disclosed on ingredient labels in the majority of product categories in the United States. Since formulations are subject to frequent alteration, alternatives might hit the market unnoticed by consumers. Therefore, in biomonitoring investigations, scientists would not always know which phthalates to assess. Many studies show that eating foods containing these molecules is the main source of exposure to DEHP and its replacements. DEHP and DINP metabolites decreased to levels five to ten times lower than pre-fasting tests in less than twenty-four hours. DINP is a commonly utilized alternative to DEHP. However, during the fasting phase, there were no discernible changes in the urine excretion of DnBP, DiBP, or BBzP metabolites, indicating that their exposures came from non-food sources. A similar attempt was made to decrease the sources of dietary exposure when households were given food packaging-free diets. Urinary DEHP metabolite levels were found to have decreased by more than 50% on average over the three-day intervention period when a standard diet was replaced with one consisting primarily of whole foods and beverages with little to no contact with plastic, aluminum, or canned packaging. Urinary DEHP was reduced by more than 90% in those with the highest pre-intervention exposures, with reductions being considerably more noticeable in these individuals. During the intervention period, there was no significant change in the levels of urine metabolites of DMP, DEP, DBP, and BBzP, indicating that the majority of exposure to these phthalates comes from non-food related sources.

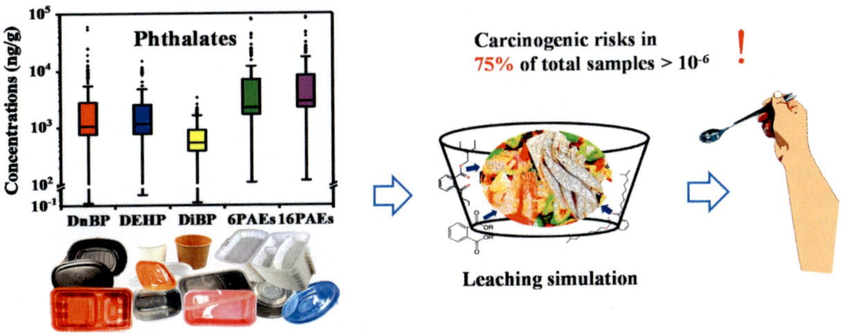

Figure 6.2. *Phthalate contamination in food.*

Source: Han, Y., Cheng, J., Tang, Z., He, Y., & Lyu, Y. (2021). Widespread occurrence of phthalates in popular take-out food containers from China and the implications for human exposure. Journal of Cleaner Production, 290, 125851. https://doi.org/10.1016/j.jclepro.2021.125851.JPG, Creative Commons License.

Phthalate levels in foods, based on limited testing, are highly variable and suggest sporadic contamination events that may occur in the source or original collection of the food, in the packaging or processing prior to market, or in the preparation leading to consumption (Figure 6.2). Meats, dairy products, and vegetable oils are examples of fatty foods that typically have greater quantities of phthalate. When phthalates with longer side chains, like DEHP, are present, phthalate esters are thought to be more lipophilic. Although the sample sizes were small, recent testing of foods bought from U.S. supermarkets revealed that different phthalates were detectable in all classes of food, with DEHP being the peak phthalate tested in most food categories, especially in pork, dairy products, vegetable oils, and grains.

This is generally in line with quantities discovered in foods from Europe, which were examined in a scenario-based model that concluded that diet had a significant impact on exposure to DEHP. Virgin olive oil from a glass container had the highest value, although the source of the phthalates in the samples was not known. The usage of PVC gaskets for metal lids on imported foods was found to be responsible for several of the highest recorded quantities of phthalates in foods, although conveyor belts, gloves, and liquid-filled tubes were other contributory sources. Food and beverage handling, processing, and packaging materials are all categorized as Food Contact Materials (FCMs) (De Tandt et al., 2021).

Consumer products include significant quantities of DEHP and BBzP. Vinyl goods include particularly high levels of DEHP, up to 40% by weight. DEP is utilized as a fragrance solvent in perfumes, lotions, hair products, and other personal care goods. It is also frequently found indoors. Conventional diapers, hand sanitizers, shaving creams, deodorants, lipsticks, automobile air fresheners, dryer sheets, nail polish, and sunscreen were also found to contain DEHP. Alternate items containing very low quantities of DEHP were also discovered, such as lipstick, hand sanitizer, and shaving cream. One of the most often found compounds in both traditional and alternative products was reported to be DEP. Four of the ten phthalates that were tested for DEP, BBzP, DEHP, and Di-n-Butyl Phthalate (DnBP) were discovered in both traditional and alternative products. Foundation cosmetics, lipstick, and perfume were among the other typical product categories that contained additional phthalates, such as Di-n-Octylphthalate (DnOP), Di-n-Hexylphthalate (DnHP), and Diisobutylphthalate (DiBP). Other substitute goods that included borax, laundry detergent, and shaving cream also contained DnOP, DINP, and di-cyclohexylphthalate. Exposure to these phthalates can occur by ingestion or inhalation of phthalates that migrate out of the product, as in the case of DEHP in vinyl curtains, as well as from direct contact with the product, such as shaving cream applied directly to the skin. The European Union (EU) has also identified DEHP, DnBP, and DiBP in sandals made of PVC and DEHP in PVC-containing sex toys as potentially important sources of exposure. Phthalates in children's toys are currently subject to various limits in the US and the EU, although high quantities have been found in PVC-based toys, with some containing as much as 40% DEHP or DINP. Exposure to inhaled clay compounds may also be significant after they have dried through baking.

6.7. OTHER TOXIC MATERIALS

The majority of Acrylonitrile-Butadiene-Styrene (ABS) plastic, which contains up to 12% of OctaBDE by weight, is made of thermoplastic resins, such as the hard plastic components of computers and televisions (Peeters et al., 2014). In addition to being added to various plastic polymers like polyvinyl chloride, polycarbonates, and High-Impact Polystyrene (HIPS), which are used in the housings of electrical and electronic equipment, as well as a coating for textiles, decaBDE is the most widely used PBDE in all markets. In order to enhance the thermal insulation of buildings, Hexabromocyclododecanes (HBCDs) are mostly added to foam made of Extruded Polystyrene (XPS) and Expandable Polystyrene (EPS). HBCDs are also applied to the HIPS housing of electrical and electronic appliances, as well as the back-coating of fabrics used in furniture. c-HBCD is prominent in commercial HBCD formulations and most abiotic samples. **Tetrabromobisphenol A** (TBBPA), the most widely used Flame Retardant (FR) in the world, is primarily utilized in epoxy resins for printed circuit boards or PCBs in computers and telecommunications equipment as a reactive flame retardant. TBBPA is covalently bonded into the polymer of goods in reactive applications. TBBPA is, nevertheless, also added to circuit boards for low-power devices like video recorders and remote controllers, as well as the plastic casing of electrical and electronic devices like printers and computer monitors. TBBPA and its metabolites have been demonstrated to be released into the environment.

6.8. CONSUMPTION PATTERNS OF THESE CONTAMINANTS

Significant variations in global consumption are reflected in BFR contamination patterns. North American citizens utilized at least 95% of the PentaBDE produced worldwide, and the population and biota of North America have the highest levels of PentaBDE in the world. TBBPA and PBDEs are the two most common BFRs in use in Asia. The second-highest volume BFR used in Europe is HBCD, after TBBPA. The demand for HBCDs has grown recently worldwide, particularly in Asia, and temporal studies indicate that human and biota a-HBCD levels are rising. Penta- and OctaBDE combinations were phased out of manufacture and use in many countries in 2005 due to health and environmental concerns. PentaBDE and OctaBDE were included in the 2009 Stockholm Convention list of persistent organic pollutants or POPs that need to be eliminated worldwide. Phase-out of DecaBDE by US producers began in 2013 after negotiations led by the U.S. Environmental Protection Agency (USEPA) (Fantke et al., 2015). The use of HBCD as a substitute for FR in textiles, plastic enclosures, and other items has expanded since the phase-out of PBDE. With special exceptions for use in building materials, the Stockholm Convention classified HBCD for global removal in May 2013. TBBPA manufacture and use are currently unrestricted, and within the past ten years, production has expanded globally, particularly in Asia. TBBPA has not gotten much regulatory attention in the United States. The European Union mandated the selective removal of polymers containing bisphenol A (BFRs), such as TBBPA, from Waste Electrical and Electronic equipment or WEEE in 2003. China, a significant global consumer of electronic debris (e-waste), approved a similar WEEE rule for BFR-treated plastics.

Large volumes of BFR-containing foam furniture, baby products, carpets, plastics, and insulation must be disposed of after their useful life to prevent future releases of PBDEs into the environment (Stubbings and Harrad, 2014).

These outdoor reservoirs include landfills, wastewater treatment plants, e-waste recycling facilities, and stockpiles of hazardous waste. House dust is the main way that people who are not exposed at work come into contact with BFRs. There is evidence that BFRs have contaminated indoor dust. Studies have shown a correlation between body loads and indoor dust concentrations, with concentrations in indoor air significantly higher than those outside. An adult U.S. resident's average exposure to PBDEs comes from house dust (82%). Dust is a direct exposure pathway through inadvertent eating, inhalation of suspended particles, and skin absorption, and it is a proxy for exposure from product usage, even if food may also contribute to PBDE exposure. BFRs are reactive to a range of polymers, including HIPS, PUF and PS foams, and epoxy resins. These polymers are used to make foam furniture, electronics, food packaging, and construction materials. Despite being widely used, there is a data gap for the majority of goods since there is a lack of information on the precise concentrations of BFRs in the polymers or final products.

6.9. PRESENCE IN HOUSEHOLD ARTICLES

Large sources of PentaBDE and DecaBDE in households are foam furniture items, such as foam Furnished Polyurethane Foam (PUF) couches and chairs, mattresses, pads, and pillows. In use and discarded furniture will continue to be a source of PentaBDE exposure for the foreseeable future because there is presently no policy in place for the identification or safe disposal of FR-treated furniture. Because they frequently put their hands in their mouths and come into contact with dust on floors and other surfaces in houses, children's exposure to BFRs is especially concerning. Infants should be especially concerned about chemical exposure from baby goods since they come into everyday contact with them at a time when their development is still quite sensitive. Children's clothing may be treated with DecaBDE, while children's foam-containing products including car seats, beds, pillows, and changing pads are sources of PentaBDEs. Sources of all commercial mixes of PBDEs include hard and soft plastic toys. Researchers assessed the FRs in PUF of 101 frequently used infant items that were bought in the United States. The items included a halogenated flame-retardant ingredient in more than 80% of them. PentaBDE congeners were found in five samples, indicating that PentaBDE-treated baby items may still be in use in the United States. Seventy to eighty percent of the toys imported into the United States are made in South China. DecaBDE was the most common PBDE in a recent investigation of toys bought in South China; it accounted for 45–79% of BFRs in toys made of hard plastic, foam, rubber/soft plastic, and plush animals. The highest total PBDE concentrations were found in hard plastic toys, with DecaBDE accounting for 65%. PBDE contents in foam toys were four to 160 times lower than in hard plastic toys. The lowest concentrations of PBDEs were seen in stuffed toys. Children may come into contact with BFRs from nylon swimwear, and stockings, all polyamide items that have been treated with DecaBDE and other FRs. Since foam cubes used in open pits intended to teach young gymnasts may contain PentaBDE at a weight percentage of 5% in the foam, children preparing for sporting events may be exposed to BFRs in gymnasiums. Numerous plastic items that come into contact with food, such as plastic wraps, packaging materials, and microwaveable containers, may expose us to BFRs. Common materials used in food packaging, polyolefin thermoplastics including Polypropylene (PP) and Polyethylene (PE), are flame-retarded with PBDEs, mainly DecaBDE. The main BFR utilized in PP products like straws, bottles, flexible yogurt containers, and microwaveable plastic food containers is decaBDE. There's a chance that butter wrapper paper contains PBDE congeners, notably DecaBDE. The primary BFR added to Thermoplastic Elastomers (TPEs) used in hot liquid tubing and fatty food containers is decaBDE. Plasticizers are also added to thermoplastic polyesters like Polyethylene Terephthalate (PET) and Polybutylene Terephthalate (PBT) to increase the flexibility of soft plastic products. Plastic water bottles and Aluminum (Al) beverage (lined) cans contain PBT and PET that have been treated with BDE-47 and DecaBDE (Shaw et al., 2014). BFRs are easily transferred from thermoplastic materials into food and drink.

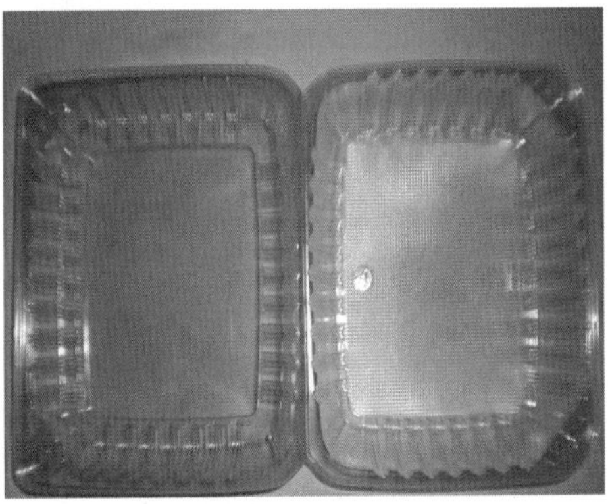

Figure 6.3. *Plastic trays.*

Source: https://commons.wikimedia.org/wiki/File:Rectangular_plastic_food_containers.JPG, Creative Commons License.

Plastic trays (Figure 6.3) used to package cheese and meat are improved with polyolefin foams, also known as rubber foam, that have been treated with DecaBDE. EPS coated with HBCDs is used to make Styrofoam, which is utilized in takeaway boxes, coffee cups, and packing for shipping. BDE-

209 and substitute FRs may be present in other food-contact goods. BDE-209 is added to toothbrush monofilaments including PBT and polyester rubber gloves. Regarding the precise concentrations of FRs in food packaging and contact materials, more data is required. Carpets, coatings, and textile materials including cotton, polyester, PP, nylon, PBT, and PET, are used to make textiles, which include drapes, carpets, furniture fabric, and back coatings. Mattresses, draperies, and tents are the main sources of exposure to PBDEs and HBCDs in the house. These materials and textile goods themselves are possible sources of these chemicals. Deca-BDE, which was once widely utilized, has been replaced by HBCD as the main BFR used in textiles in the United States. DecaBDE is the second most common BFR applied to textiles in Japan, after HBCDs. On the estimated breakdown and emission rates of BFRs from home textiles, there is a paucity of data. PBDE emission rates were lower than HBCD emission rates from curtains, suggesting that HBCDs more easily seep into indoor environments from textiles. When exposed to sunlight, HBCD exhibited greater resistance to photodegradation than BDE-209; however, BDE-209 added to the curtain material photodegraded and contributed to the synthesis of Polybrominated Dibenzofurans (PBDFs), indicating that treated products used daily may be sources of PBDFs in indoor dust and air.

Researchers carried out an extensive examination of BFRs in consumer goods, encompassing various parts of Japanese laptop computers and LCD TVs. In the majority of the TV and laptop computer components examined, decaBDE accounted for 73–98% of the total PBDE concentration, making it the predominant congener. The plastic covers on the back and front of LCD TVs had the highest RPBDE concentrations, followed by printed circuit boards and LCD panels. The cooling fans and speakers in laptops had the highest RPBDE concentrations, followed by the LCD panels and keyboards. The printed circuit boards and front covers of LCD televisions had the greatest concentrations of HBCD. TBBPA concentrations varied widely, from 0.007 mg/kg in LCD TV panels to 9,500 mg/kg in laptop cooling fans and speakers, extremely high levels. The two main BFRs utilized in building materials and housing insulation are HBCDs and PBDEs, yet there is a dearth of information on the application of these compounds in goods, including concentration data. The two main flame retardants utilized in EPS and XPS, high density/flexible, and rigid PUF, have been HBCDs and PBDEs. The most common BFR found in PUF, PE, EPS, and XPS put in insulation foam boards is HBCD. PentaBDE can be found in high density, flexible PUF, whereas BDE-209 and other FRs can be found in stiff PUF. A common environmental matrix used to track BFR contamination in indoor contexts is house dust. Dust from the UK was found to have the highest BDE-209 values (Bramwell et al., 2017). House dust samples from the United States and other nations had lower BDE-209 concentrations, which may be attributed to the UK's larger use of DecaBDE. Dust from an e-waste location in Vietnam has comparatively high BDE-209 concentrations. International variations in indoor HBCD contamination may occur, although they are not significant. As expected, given their greater use of HBCD, Japan and the UK have the highest concentrations of HBCDs in dust. Human body HBCD concentrations in dust have been linked, indicating that dust is a significant factor in determining human exposure to HBCDs. There is little information available regarding TBBPA levels in indoor dust. In general, TBBPA concentrations are lower than those of PBDEs and HBCDs, which is in line with the compound's principal use as a reactive flame retardant. The age of the electronics was shown to have a strong correlation with dust concentrations, indicating that newer electronics might contain less TBBPA than older models. TBBPA concentrations in dust samples from Japanese households in 2006 were found to be twice as high. Dust from surfaces of computers, microwaves, refrigerators, printers, and other devices has been shown to contain higher quantities of TBBPA. Dust swabs from refrigerators, microwaves, and CRT computer monitors had the highest amounts. Computer tower and copier dust had the lowest concentrations. Similarly, substantial quantities were found in Japanese PC boards for electronics.

6.10. ENGINEERED NANOMATERIALS

Although (ENMs) have special features that can be used for a variety of beneficial purposes, there are uncertainties about whether ENMs could be harmful to human health. The possibility of ENM migration into food and the possible toxicity of such released ENMs are the main topics of interest for evaluating the safety of nanocomposite Food Contact Materials (FCM). Due to the quantum confinement and the high degree of surface area compared to macro-scale versions of the same fundamental substance, matter with particles on the nanoscale exhibits distinct chemical and physical properties. Therefore, it is anticipated that ENMs will be used in a wide range of industries, such as electronics, biomedicine, and construction, in addition to consumer goods including food packaging, sports equipment, cosmetics, and nutritional supplements. In addition to the potential advantages of these uses, there are uncertainties about potential health and environmental hazards from the potential implementation of nanotechnology in the future. Any new technology used in the manufacture or packing of food contains some risks that should be assessed; nevertheless, assessing the toxicological characteristics and risk of exposure to ENMs presents special methodological difficulties not seen in other chemical substances. The ways that ENMs may interact with biological systems are unknown because research on the chemical and physical properties of ENMs is still ongoing. After being approved for usage, ENMs can be applied to food or integrated into FCMs (packaging) for a variety of uses. Foods that are specifically designed to have distinct size-dependent qualities are referred to here since they naturally contain a variety of nanoscale structures.

Research on the safety of Food Contact Materials (FCMs) containing ENMs has been given top emphasis since ENM-based FCMs may be closer to commercialization than ENMs as direct food additives (Pillai et al., 2014). Compared to food packaging made of paper, glass, or metal, those made entirely of plastic polymers have some advantages in terms of functionality and cost. Polymers are inexpensive and lightweight. They are easily formed into a wide range of shapes and may be processed at low temperatures. Plastic containers are flexible and strong enough to withstand impacts (drops) that could break glass containers or permanently distort cans. Additionally, a wide range of physical characteristics of synthetic polymers can be precisely adjusted, such as the degree of moisture or oxygen exchange. Moreover, additives can be added to the polymer matrix to impart additional features that are challenging to obtain by polymer processing alone, or polymers can be blended to create mixes with intermediate properties. This can include more complex elements like antimicrobials or compounds that can react to the presence of infections, chemical pollutants, or changes in environmental conditions, as well as more basic additives like pigments or UV light blockers.

6.11. OTHER SYNTHETIC POLYMERS

There are several polymers used in food packaging. Natural resins and waxes were the first instances in history, but synthetic polymers have mostly replaced them since they were prone to breakdown and other chemical reactions. One of the earliest types of polymers to be used extensively was polyolefin, which is defined by continuous chains of carbon–carbon single bonds. Polyolefins are still frequently used today. The most common polymer in this class, Polyethylene (PE), is used in a wide range of products, including ice cream containers, plastic sandwich bags and carton liners. PE is created in a variety of forms, all of which are highly non-polar materials with minimal moisture permeability and are mainly differentiated by their relative densities. Other physical characteristics, such as hardness, melting temperature, and crystallinity, might differ significantly according to processing, molecular, and density parameters. Polystyrene (PS) and Polypropylene (PP), two substituted polyolefins, are commonly used in food packaging for certain applications that require higher heat resistance or where the various grades of PE are not strong or hard enough, or where certain types of processing are necessary. Due to its superior thermoforming qualities,

Polyvinyl Chloride (PVC) is another substituted polyolefin that is widely utilized in food packaging; however, its application is presently decreasing. Due to their high crystallinity, packaging materials containing PE copolymers, like Ethylene Vinyl Alcohol (EVOH), have become increasingly popular as substitutes for metal and glass containers. EVOH provides strong barriers against the migration of other nonpolar volatile organics, like flavor compounds, and the diffusion of oxygen. Polyesters and polyamides are examples of non-polyolefinic polymers that are widely employed. Among these materials, Polyethylene Terephthalate (PET) is particularly well-liked for usage in carbonated beverage containers and single-use water bottles because of its excellent optical transparency, strong durability, and resistance to breaking. Because of their biodegradability, bio-derived and bio-inspired polymers like thermoplastic starch, Polylactic Acid (PLA), and others are currently being developed as alternatives to petroleum-based polymers.

Although synthetic polymers have transformed the food sector, they also have a variety of drawbacks that may limit their applicability or, in certain situations, raise questions about their safety. These drawbacks include chemical stability and the difficulty/cost of recycling or disposing of the material; however, the main functional issue with plastics is their permeability, which is usually significantly higher than that of glass or metal. Polymers have the capacity to absorb tastes, colors, and odors from food due to their increased permeability. Additionally, it lessens the capacity of polymer films to shield food that has been stored from oxygen and water vapor, two things that are known to be the main causes of food spoiling. In addition to affecting food quality, polymer permeability lets components that are added to the polymer during synthesis move into the environment or the food matrix at different stages of the product's life. Intentional additives like plasticizers, pigments, antistatic agents, foaming agents, and antioxidants or antiaging substances to stop polymer degradation in the presence of light, oxygen, and free radicals are examples of substances that are added to polymers during fabrication or processing. Although the permeability of these materials can be reduced through polymer processing and engineering, there is concern that any novel polymer addition or material that comes into contact with food could provide a migration risk. These uncertainties extend to nanoscale additives as well, since these substances are being created for a variety of specific food contact applications and there are a lot of unknowns around their ability to transfer from plastics into consumer food. The main factors that cause food to deteriorate are oxygen or water vapor found in the uncontrolled outdoor environment. One of the main purposes of a food packaging material is to prevent large-scale gas transfer from the exterior to the inside of the package. Because the structures of metal and glass are made up of vast, three-dimensional networks of tightly bound atoms, their permeability to gases is almost negligible. Due to the relative porosity of most organic polymers, which can occasionally even reach large molecules, polymer permeability in food contact applications poses a serious engineering issue. The total rate at which a gas can pass through a standard thin polymer film is determined by the processing parameters as well as the molecular structure of the polymer. The gas will go from the high to the low concentration zone through a thin film of uniform thickness that divides the region of high gas concentration from the region of low concentration.

It has long been known that elemental and ionic silver molecules have antibacterial qualities (Hamad et al., 2020). However, it has also been found that nanoparticulate silver has the ability to either kill or inactivate a variety of bacteria, viruses, fungi, and other microscopic organisms. Whether the mechanism of action or antibacterial activity of silver nanoparticles (AgNPs) differs from that of free silver ions is still up for dispute. AgNPs are easily integrated into polymer matrices to create polymeric products that prevent the growth of microorganisms on their surface. AgNP-based PNCs, for instance, have been shown to prevent the growth of methicillin-resistant Staphylococcus aureus in urinary catheters and hospital dressings. AgNPs have been used in food packaging and processing materials such as cellulose pads meant to be used for fresh food and meat storage as protection against bacteria and yeast, LDPE-based plastic bags to reduce fruit decay, edible alginate coatings for fresh produce, and Low-Density Polyethylene (LDPE) multilayer films with efficacy against the fungus Aspergillus niger. The researchers demonstrated that the use of packaging materials incorporating nanosilver has the potential to enhance both the safety and quality of packaged foods by reporting that AgNP-PNC coated produce had higher consumer acceptance than controls (uncoated vegetables). In PNC materials intended for food packaging, many classes of Metal Oxide Nanoparticles (MONPs) have also been included as active elements due to their demonstrated antibacterial qualities. MONPs consisting of titanium dioxide (TiO2), zinc oxide, magnesium oxide, copper oxide, and other oxides have been demonstrated to possess antibacterial characteristics. Since TiO2 particles have a unique photoactivated mechanism that causes the absorption of UV light to stimulate the production of ROS or reactive oxygen species, which in turn breaks cellular membranes near the nanoparticle, they have been identified as particularly promising antimicrobial materials.

Chapter 7

URBAN MANAGEMENT AND MICROPLASTICS

Contents

Introduction 140
7.1. Distribution of Microplastics 141
7.2. Wastewater Treatment Plants 142
7.3. Activated Sludge Process 144
7.4. Sequencing Batch Reactor 145
7.5. Processing of Microplastics 146
7.6. Bacterial Dominance Patterns 147
7.7. Nitrogen Bacterial Dynamics 148
7.8. Changes In Microbes Due to Bioplastics 149
7.9. Research Findings 150
7.10. Inhibitory Effects of Polymers 152
7.11. Biological Treatment 153
7.12. Monitoring 154
7.13. Methane and Plastics 156
7.14. Microplastics Alter the Bacterial Abundance 158
7.15. Treating Microplastics in Wastewater 159
7.16. Mechanisms of Microplastic Damage 162

INTRODUCTION

Although the procedures are not yet standardized, the three basic processes of microplastics analysis in an urban water system are sample collection, sample purification, and sample identification (quantitative analysis) (Schymanski et al., 2021). Since microplastics are commonly found in a variety of urban water sources, including freshwater, sediment, wastewater, sewage sludge, and drinking water, samples from urban water may be collected using several methods. Furthermore, most research uses quality control to prevent any discrepancies owing to sample contamination and loss. When microplastics (3–5 mm) are directly visible on the surface of sediments, selective sampling is frequently employed to capture them. Bulk sampling, which is typically used for biological and sediment samples, involves bringing all samples to the lab without reducing their number. Since water samples frequently have low levels of microplastics, substantial sample volumes are required for examination. Concentration in the field lowers the sample volume, which is advantageous for transportation. Plastic materials are used in daily life, and as demand for them grows, the annual production of plastic materials is growing exponentially. More plastic has been produced in the last 70 years than any other industrial product. The annual production was recorded to have increased in comparison to the initial large-scale plastic manufacture in the 1950s. The amount of plastic produced increased from 1.7 million tonnes (MTs) in the 1950s–368 MTs in 2019. In the 1990s, it surpassed 100 MTs.

It is anticipated that this increase will continue, with yearly production estimated to double (600 MTs) by 2025 and triple (>1,000 MTs) by 2050, taking into account the present pace of plastic consumption and population growth. 6,300 MTs of plastic refuse were produced in 2015 alone, and 8,300 MTs have been produced thus far (Norton, 2023). Additionally, it was shown that just 9% of the plastic manufactured was recycled, 12% was burned, and 79% ended up in the environment or landfills. It is estimated that 12,000 MTs of plastic refuse will be present in landfills and the environment by the year 2050 (Geyer et al., 2017). Many different plastic materials are released into the environment as a result of the high manufacturing volume and wide range of applications. Large volumes of plastic waste are thereby contributing to a number of environmental issues, including landfill saturation and marine pollution. Microplastic contamination is an excellent illustration of the pollution that comes from plastic waste. Today, tiny pieces of plastic that have weathered for many years and are difficult for biodegradation can be found all over the world. These, also known as microplastics, can be discovered in freshwater and saltwater, among other aquatic settings. Tap and bottled water have also been reported to contain microplastics.

7.1. DISTRIBUTION OF MICROPLASTICS

Microplastics have also been found in food, soil, and air, in addition to the aquatic environment. There is not enough data to make judgments on the toxicity of microplastics. Microplastics may, however, have detrimental effects on health. These effects include the buildup of microplastics in organs, oxidative stresses, inflammation, and changes in the expression of genes linked to stress in different organisms. Additionally, it has been shown that microplastics act as chemical transporters, hastening the environmental movement of heavy metals and organic contaminants. Therefore, one of the biggest issues facing the environment and public health is the widespread microplastic contamination that is becoming a problem. Therefore, it is crucial to comprehend the entire process by which microplastics are produced as well as the environmental compartments into which they enter. Furthermore, there is an immediate need for research on methods to reduce contamination and remove microplastics. Primary and secondary microplastics are two categories of microplastics. Primary microplastics are synthetic particles that fall into the microplastic size category. Examples of these particles include nano-sized latex plastics found in paints and coatings, microbeads found in liquid soaps and cosmetics, and abrasives found in cleaning agents. Primary microplastics are discovered in natural and urban environments as a result of the mass production and use of consumer products containing microplastics. Due to their slow breakdown, primary microplastics are becoming a global problem. Plastics may weather into tiny fragments when they are exposed to the natural environment. Secondary microplastics are small fragments of plastic that have broken down in this manner. Any plastic that enters the environment, such as plastic bags, containers, fishing nets, and water bottles, is a source of secondary microplastics. The physicochemical characteristics of the plastic product, the rate and duration of weathering (photodegradation), and environmental conditions specific to the area are likely to influence the occurrence rate of secondary microplastics (Fotopoulou and Karapanagioti, 2019). For instance, physicochemical changes in plastic surfaces subjected to high temperatures and intense UV radiation over extended periods may cause the plastic material to become brittle, making it more prone to fragmentation and decomposition. There is less information available on how quickly microplastics break down and degrade in drinking water.

Reports of microplastics originating from the water treatment process have also been raised. The ozone treatment process effluent included more microplastics than the influent water did. The reason for this is that shearing forces break plastics into smaller pieces, which increases the quantity of microplastics. In addition, the effluent of water treatment facilities occasionally included higher quantities of Polyacrylamide (PAM) than normal water. This is due to the coagulation process using a coagulant that contains PAM. Microplastics are an issue not only in the coagulation and ozonation processes but also in the membrane filtering procedure. Membrane filtration systems are currently utilized in water treatment facilities, and it is acknowledged that using a membrane filter is an efficient way to remove microplastics from water. Backwashing membrane filters in between operations is necessary to keep them clean and free of blockage. An area of the membrane that has been in use for a long time may become damaged due to the high pressure and air from backwashing, which increases the likelihood that a portion of the membrane surface may break into tiny pieces. Furthermore, oxidizing environments created by chemical agents like NaOH, H_2O_2, and NaClO cause the membrane to become more brittle and enable the dissolution of materials like PVC. Microplastics may be released from organic membrane filters in water treatment facilities as a result of physical and chemical agents, and mechanical stress. When old membrane filter systems that include sealing, casing, gaskets, etc. can also release microplastics through extended mechanical weathering (Tian et al., 2023).

Figure 7.1. *Wastewater treatment plants.*

Source: https://en.wikipedia.org/wiki/Sewage_treatment#/media/File:Moscow_Kuryanovo_wastewater_plant_asv2018-08.jpg, Creative Commons License.

7.2. WASTEWATER TREATMENT PLANTS

The majority of plastic microbeads are deposited in wastewater treatment plants (Figure 7.1) after entering the sewage system through rainwater or domestic wastewater. Due to their size, most of them are eventually turned into sludge or released into rivers containing effluent because standard wastewater treatment methods are unable to reject them effectively. Over 90% of microplastics that enter wastewater treatment plants often end up in sludge and the aquatic environment. A significant buildup of microplastics brought about by the release of microplastics into the aquatic system has created pressure. One of the primary sources of pollution with microplastics is the effluent from wastewater treatment plants or WWTPs. These may also serve as the primary treatment facility for reducing the amount of microplastics that pollute the aquatic ecosystem. It is crucial to provide an overview of the destiny of microplastics in wastewater treatment facilities. Gaining further insight into the effectiveness of how conventional wastewater treatment systems remove microplastic could lead to recommendations for enhancing the removal of microplastics and reducing their release into natural waterways. There are regional or national variations in the amount of microplastic present in WWTPs. Since there has been very little information released regarding the microplastic contamination in industrial WWTPs, the majority of the WWTPs that have been known are generally domestic WWTPs.

By utilizing various post-treatment procedures, researchers examined the removal of microplastics from municipal wastewater treatment plants. Their findings suggested that 99.9% of the microplastics could be removed using a membrane bioreactor. Some researchers showed that drinking water treatment using a membrane bioreactor could remove microplastics from contaminated surface water. This was mostly due to the membrane's rejection of the microplastics. Scientists looked at how microplastic buildup affected the removal of microplastics from wastewater using a membrane bioreactor system. In a membrane bioreactor, microbial growth was suppressed by Polypropylene (PP) microplastics at a concentration of 0.14–0.30 g/L. By continuously scrubbing the membrane and increasing the diversity of the microbial life, the PP microplastics reduced membrane fouling. Elevated PP concentrations also decreased proteobacteria and increased the number of clostridia. Similar findings were also achieved by many researchers showing that the buildup of microplastics in a membrane bioreactor system may influence microbial diversity and reduce membrane fouling during long-term operation. The use of membrane bioreactors is dependent on low-cost membrane and technological advancement. Cost-effectiveness analyses of several strategies for removing microplastics from wastewater treatment systems were carried out. According to some estimates, the membrane bioreactor method reduced the annual release of microplastics into the water by billions of particles, making it the most cost-effective technology. Combining MBR with sludge incineration could lessen the environmental impact by reducing the possibility of microplastic contamination of soils from sludge usage. Raw sewage from wastewater treatment facilities has regularly been found to include microplastics. Because of their size, microplastics are predicted to present a substantial challenge to the current design of wastewater treatment systems, particularly biological processes. Early in primary and secondary treatments, it has been discovered that a sizable fraction of microplastics can be eliminated by adsorption to solid surfaces or sludge. Small microplastics may hinder the natural biodegradation process of organic pollutants and prevent sludge bacteria from growing in the Activated Sludge Process (ASP). Microplastics may have detrimental impacts on biological treatment processes (Wei et al., 2021), such as changing the composition of microbial communities, preventing microbial growth and reproduction, and increasing the adsorption of contaminants. These procedures included the Membrane Bioreactor (MBR), anaerobic-anoxic-aerobic (A_2O) process, Sequencing Batch Reactor (SBR), biological filter, Moving Bed Biofilm Reactor (MBBR), and ASP.

7.3. ACTIVATED SLUDGE PROCESS

In ASP, organic contaminants in wastewater are consumed by microorganisms that thrive in the aeration tank under aerobic conditions. Extracellular Polymeric Substances (EPS) are secreted by microorganisms to absorb organic materials and nutrients, including phosphorus and nitrogen. This mechanism also breaks down certain pollutants that bacteria can readily access. Because some of the microplastics are entrapped by the biological flocs and form a biofilm on their surfaces, microplastics may be more likely to settle in ASP sludge. Regardless of form, the biological flocs captured microplastics, but ASP sludge included a sizable accumulation of microplastics in fiber and fragment forms. The majority of microplastics found in secondary sludge (50–100%) were found in fibers, then fragments (5.7–25%). Fibrous microplastics may become stuck to the surface of biological flocs as a result of aeration, which promotes the growth of microorganisms, and ultimately settle in the sedimentation tank. The ASP sludge contained a substantial amount of fibrous microplastics, and the ASP obtained a comparatively high removal efficiency. Microplastics in wastewater have the potential to worsen the treatment performance in ASP and raise the content of newly emerging organic contaminants such as phthalates and bisphenol A (BPA). The association between EPS and Polystyrene (PS) nanoplastics may change as a result of the increased concentration of microplastics in the ASP, which could restrict microbiological development. When microplastics stick to microbial cells, they can cause cell dysfunction.

There were effects on denitrification but minor inhibitory effects on nitrification when there was a high concentration of microplastics. Nonetheless, the polymer types of the microplastics had a significant influence on the exhibitory or inhibitory effects on the treatment performance. Denitrification was impeded and nitrous oxide emission was suppressed during nitrification due to the buildup of hazardous chemicals and a high concentration of PVC microplastics in the tank. The major objectives of other suspended-growth processes, such as the oxidation ditch and A2O process, are to convert organic materials into stable solids, carbon dioxide, or other stable nontoxic chemicals, and to decrease Biochemical Oxygen Demand (BOD) to acceptable quantities (Jahan et al., 2012). Up to 95.6% of the microplastics were removed by the oxidation ditch, which has been considered a reasonably successful unit operation; of these, 92.3% were in the form of fibers. The adhesion of fibrous microplastics on the surface of biological flocs may also be the cause of this outcome. Because microplastics may collect during sludge recirculation, a relatively low clearance of microplastics was observed in the A2O method. Additionally, there was a 22.1–257.1% rise in the abundance of fibrous microplastics and a 115.8% increase in those with diameters 0.1–1 mm. Stirring and aerating wastewater in tanks can both raise the friction force inside the wastewater, which can lead to microplastic fragmentation and an increase in the number of microscopic fibrous microplastics.

7.4. SEQUENCING BATCH REACTOR

Another common suspended-growth process used in wastewater treatment facilities is called the Sequencing Batch Reactor (SBR). Over 99% of the microplastics in the wastewater were eliminated with SBR. While some research suggested that microplastics might have an impact on SBR, no discernible effect was seen. Various dosages of Polyamide (PA), PE, PS, and PVC microplastics did not affect SBR treatment performance; some researchers found no effects of microplastics on SBR. Regardless of the type of polymer, the microplastics at a concentration of 1 mg/L could stimulate the secretion of EPS, and their presence did not affect the efficiency of ammonia oxidation. The standard methods of attached-growth processes in biological wastewater treatment worldwide include trickling filters, Submerged Aerated Filters (SAF), rotating biological contactors, and Biological Active Filters (BAF). The quantity of fiber and fragment microplastics rose by 4.7% and 210.9%, respectively, as a result of possible microplastic fragmentation in BAF. Microplastics may be adsorbed on the filter surfaces in various growth processes, keeping them in the sludge. Nonetheless, the fate of microplastics in attached-growth processes has not been extensively studied. The impact of microplastics on the treatment efficacy of various biological filter systems requires investigation. Advanced wastewater treatment systems like the **Membrane Biofilm Reactor (MBR) and The Moving Bed Bioreactor (MBBR)** can further remove nutrients, suspended particles, and BOD from wastewater. ASP removes nutrients and organic matter from wastewater during the MBR process, while membrane technology efficiently lowers the amount of microorganisms and suspended solids in the water. Microfiltration membranes measuring between 0.1 and 50 μm and ultrafiltration membranes measuring between 0.001 and 0.1 μm are typically used in the MBR process (Arévalo et al., 2012). Microplastics are directly caught in the process by membranes with extremely small pore sizes. Additionally, it has been demonstrated that while a small percentage of fibrous microplastics might pass through the membrane, 99.9% of the microplastics in wastewater could be caught by the membranes.

7.5. PROCESSING OF MICROPLASTICS

Fibers and pieces made up 88.3% and 11.7%, respectively, of the microplastic in the MBR sludge. In MBR sludge, researchers found microplastics in the forms of fragments (75%) and films (25%), but no fibrous microplastics. It is implied that the removal in MBR is a function of size but not formed by the variance in shape composition of microplastics in MBR sludge. Furthermore, total removal of microplastics is possible when an ultrafiltration membrane with a nominal size of 0.2 μm is used. Additionally, MBR has been considered the best method for removing microplastics when the membrane pore size is less than 0.1 μm. Microorganisms proliferate on freely moving carriers in the reactor during the MBBR process. More research is needed in this area because the fates of microplastics in MBBR have not been adequately explained. There hasn't been much research done on how microplastics affect sophisticated wastewater treatment systems. In comparison to the control and low dosage groups, the MBR's high dosage group had a greater removal efficiency of ammonia. The efficacy of biological wastewater treatment as well as the effectiveness of nitrification and denitrification were considerably impacted by the presence of microplastics. It investigated how microplastics affected the MBR process in terms of floc sizes, surface characteristics, and chemical compositions. Reactors with a high dosage of microplastics had a substantially different microbial diversity than the control. Therefore, the microbial diversity and structure within sludge flocs are significantly impacted by the buildup of microplastics. Differently shaped microplastics might affect sludge bacteria differently. Due to their adhesion and ability to cause cell membrane damage, fibrous microplastics may have negative impacts. Nevertheless, fragmented microplastics have the potential to infiltrate cells and hinder the growth of microorganisms. In biological treatment systems, EPS made up of proteins, polysaccharides, lipids, and humic acid can influence sludge aggregation. Microplastics could encourage the formation of sludge.

Bacteria, protozoa, and fungi are among the many microorganisms that make up Activated Sludge (AS), which is used in wastewater treatment operations to extract organic materials from wastewater. The primary roles of AS, supported by nitrite-oxidizing bacteria, ammonia-oxidizing bacteria, and ammonia-oxidizing archaea, are nitrification and denitrification. Exposure to PA and PVC microplastics at a dosage of 1 mg/L for 14 days hindered the nitrification of AS. The PA and PVC microplastic densities are comparatively higher than those of the reactor's effluent, which is consistent with their settling in AS. These microplastics accumulated and became more concentrated in the tanks, giving them a chance to come into contact with nitrifying sludge. The addition of PA and PVC microplastics at a concentration of 1 mg/L altered the microbiological morphology of the sludge.

The spherical shape of the nitrifying sludge is thought to be connected to the friction it experiences with PA and PVC microplastics (Wolff et al., 2021). Methane production can be stimulated by PVC microplastics at a concentration of 10 particles/g, but it can be inhibited by PVC microplastics at higher concentrations. Mechanistic investigations revealed that the main cause of the lower methane generation was BPA leaching from PVC microplastics, which had a considerable inhibitory influence on the hydrolysis-acidification process. The microbial community was shown to have changed away from methanation and hydrolysis-acidification as a result of the long-term effects of PVC microplastics. When antibiotic-resistant bacteria enter the environment, high cell density can act as a protected breeding ground, promoting their growth and, over time, the horizontal gene transfer of antibiotic-resistant genes across the environment.

7.6. BACTERIAL DOMINANCE PATTERNS

Heterotrophic bacteria obtain their energy and carbon from organic substances, which they use for synthesis. Through colonization on the surface of microplastics, microbes use the organic carbon present in the microplastics as their source. The strategy used by microbes to survive in difficult environments during early colonization is called biofilm formation, which is started by planktonic bacteria through mutual cohesion.

Following a period of time during which physical pressures are applied, the attachment becomes immobile and is permanently adsorbed on the surface, causing the formation of microcolonies and the production of EPS. There are three different phyla of Proteobacteria: Alphaproteobacteria, Betaproteobacteria, and Gamma proteobacteria. The two most common heterotrophic bacteria on the surface of microplastics in the form of biofilms are Bacteriodetes and

Actinobacteria. The dominance of Alphaproteobacteria indicated that other bacteria could colonize during the early stages. Bacteriodetes served as biofilm conditioners by breaking down dissolved organic matter, while Actinobacteria assisted in the digestion of organic matter.

7.7. NITROGEN BACTERIAL DYNAMICS

One of the main bacterial groups that aid in nitrification and denitrification is ammonia-oxidizing bacteria. Depending on their quantity, size, and type of polymer, microplastics can either improve or worsen the efficacy of nitrification and denitrification by blocking ammonia-oxidizing bacteria. Ammonia oxidized entirely and at a slower rate when there was a low abundance of PA microplastics (50–100 mg/L). Consequently, there was very little effect on microorganisms that oxidize ammonia. However, the activity of ammonia-oxidizing bacteria was suppressed when the dosage of PA microplastic reached 200 mg/L. As a result, partial oxidation was discovered, with residual ammonia at 7.6 mg/L. The ammonia oxidation rate was marginally increased by PP and PVC when 1000 particles/L of microplastics were administered. However, when the dosage was increased to 10,000 particles/L, the performance declined. While polyester, PE, and PS all hindered ammonia oxidation, the negative effects were marginally mitigated when the dosage was increased from 1000 to 10,000 particles/L. The addition of PVC microplastics significantly reduced the ammonia-oxidizing bacterial activity. When the dosage of PVC microplastics was increased to 1000–10,000 particles/L, the baseline ammonia oxidation rate dropped by 28.2 to 50.0% (L. Li et al., 2020). The release of hazardous chemicals from PVC microplastics is attributed to the suppression of ammonia-oxidizing bacteria.

By changing the nitrification genes, the quantity and polymer types of microplastics reduced the efficiency of nitrification. The abundance of ammonia-oxidizing bacteria was greatly reduced by the presence of PA, followed by PE, PS, and PVC at a fixed dosage of 1 mg/L of microplastics. Even though the number of ammonia-oxidizing bacteria decreased in the initial stages, their activity level for ammonia oxidation remained high. Following a period of acclimatization, the microorganisms started growing again at the same time, and the quantity of ammonia-oxidizing bacteria increased as a result of their adaptation to the presence of PA microplastics at a concentration of 50 mg/L. Nitrite-oxidizing bacteria are another type of bacteria that cause nitrification and denitrification, much like ammonia-oxidizing bacteria. Compared to ammonia-oxidizing bacteria, whose specific nitrite oxidation rate remained below 2.0 mg N/g, the impact of PA microplastics on the activity of nitrite-oxidizing bacteria was less pronounced. The rate of nitrate generation was decreased by microplastics made of PE, polyester, PP, PS, and PVC.

Regardless of their quantity, microplastics reduced the nitrite-oxidizing activity during nitrification. Since ammonia-oxidizing bacteria grew 2.6 times faster than nitrite-oxidizing bacteria and because they were the predominant strains in the partial nitrification process, the inhibition of nitrite-oxidizing bacteria was reduced with 1000 particles/L of PVC microplastics.

Just like ammonia oxidizing bacteria, the types of polymers and the quantity of microplastics had an impact on the nitrification genes. The addition of PS, PE, and PA considerably increased the abundance of nitrite-oxidizing bacteria with the fixed dosage of microplastics at 1 mg/L. The increase in nitrite-oxidizing bacteria abundance by 4.7% was marginally affected by the presence of PVC. The growth of nitrite-oxidizing bacteria may be inhibited by the hazardous chemicals generated from PVC microplastics. The quantity of nitrite-oxidizing bacteria increased by 286.7%, 397.5%, 187.3%, and 60.1%, respectively, when the doses of PS, PE, PA, and PVC microplastics reached 100 mg/L. These findings showed that when the treated microplastics reduced the growth of ammonia-oxidizing bacteria but promoted the abundance of nitrite-oxidizing bacteria, the abundance of nitrite-oxidizing bacteria was negatively linked with the abundant of ammonia-oxidizing bacteria. It's interesting to see that nitrite-oxidizing bacteria (Nitrospira) abundance was unaffected by PA microplastics. They did, however, have little effect on the overall decrease in bacterial abundance, with ammonia-oxidizing bacteria being the primary source of the effects on partial nitrification. The quantity and activity of nitrite-oxidizing bacteria may not be threatened by the presence of microplastics.

7.8. CHANGES IN MICROBES DUE TO BIOPLASTICS

The adhesion of microplastics to the surface of cell membranes lowered EPS protective barriers and enhanced contact between microplastics and important enzymes. Microplastics may be more likely to get through the spaces created by the chains of biopolymers. This penetration may increase the possibility of interactions between proteins, phospholipids, and microplastics, which could lead to cell-associated toxicity and the elimination of enzymatic activity.

The production of oxidative stress at the cellular level due to the toxicity of leached compounds from microplastics could be another factor for the decrease of enzymes (Li et al., 2024). Microplastics may produce Reactive Oxygen Species (ROS) in cells, which can lead to lipid peroxidation, membrane damage, and cell deformation. This damage prevents electron transmission and impairs energy exchange both within and outside the cell. Because there were only so many bacterial enzyme molecules available, the cell could not eliminate the excess ROS and eventually died when the concentration of ROS surpassed the capacity of the cell to repair itself. Microplastics have the potential to negatively impact enzymatic activity because they physically obstruct the movement of electrons and energy within and outside of cells, causing the generation of ROS, or both. Crucial elements influencing microbial aggregation in biological treatment are the properties and elements of EPS. EPS are necessary to preserve the structure of sludge flocs and shield the microbial community from newly developing organic pollutants. On the other hand, in biological treatment, microplastics and newly discovered organic pollutants had inhibitory or synergistic effects on EPS secretion. Different effects on EPS can be caused by the treatment of microplastics and the length of time a specific polymer type is exposed.

PVC microplastics can be briefly exposed to increase EPS secretion and improve wastewater phosphate removal. On the other hand, over time, PVC microplastics increase cell death by dissolving the structure of sludge flocs. On the other hand, bacteria may respond differently to exposure to different types of polymers. A brief exposure to PA66 microplastics damaged the microorganisms in the sludge and hindered their ability to proliferate. Over time, the sludge's microbes acclimated to the presence of PA66 microplastics, which increased the sludge's production of signaling molecules and encouraged the secretion of EPS. EPS may respond differently to exposure to various doses of microplastic. When exposed to high doses of PET microplastics, EPS synthesis was reduced, whereas EPS secretion was boosted when exposed to low levels of PET microplastics. However, by creating a barrier from loosely bound EPS and producing more protein from securely bound EPS, microplastics may shield enzymes from harmful environments. The toxicity released by polyether sulfone microplastics and BPA, as well as the contact between bacteria in anaerobic granular sludge, may be eliminated by the tryptophan-like proteins in loosely bound EPS. The presence of BPA and microplastics may promote the synthesis of EPS that is firmly bound, increasing the protein content and preventing the harmful leachate of microplastics (Lin et al., 2020).

7.9. RESEARCH FINDINGS

Research on the possible effects of microplastics on sludge dewaterability has been limited up to this point. The capillary suction time test was carried out using filtration. The outcomes showed that microplastics had a major impact on the dewaterability of sludge. The capillary suction time and the quantity of PS microplastics were shown to be negatively correlated. A 17.0% reduction in capillary suction time was observed upon increasing the concentrations of PS microplastics from 0–10 mg/L. This suggests that sludge dewaterability was aided by a low abundance of small PS microplastics. However, the increase in capillary suction time indicated that the sludge's dewaterability was hampered when the concentration of small PS microplastics reached 100 mg/L. The high concentration of microplastics may change the distribution and composition of EPS by reducing the activity of important bacteria and inhibiting their action. Additional investigation is required to elucidate the potential pathways. The capillary suction time increased steadily by 29.9% from 0–300 mg/L as a result of the rise in the quantity of big PS microplastics. The size of the microplastics may be crucial to sludge dewatering, since the large microplastics impede sludge dewaterability. One potential mechanism for sludge dewatering degradation could be the physical crushing of sludge flocs. Additionally, comparable to PS microplastics, PE microplastics degraded sludge dewaterability by about 30% when taking into account variations in polymer types of the same size. PVC microplastics, on the other hand, had a negative impact on the dewaterability of sludge. PVC may produce harmful substances that disrupt the microbial community structure. The relationships between the characteristics of microplastics and the dewaterability of sludge necessitate more research into the causes.

Microplastics are not the intended contaminants to be eliminated in the secondary treatment because the primary goal of biological treatment is to decrease suspended particles and residual organics in the primary effluent. The role of secondary sedimentation and the redistribution from back flow are two aspects of the microplastics' secondary treatment pathways that have not been thoroughly investigated. The availability of membranes, rather than interactions with microorganisms in activated sludge, is a primary factor in trapping different shapes and sizes of microplastics, even in MBR effluent when the concentration of microplastics is very low. However, secondary sedimentation's contribution to the elimination of microplastics is disregarded. The majority of studies have used the variations in microplastic abundance between primary and secondary effluent to calculate the elimination effectiveness of microplastics in secondary treatment. By settling the biofilm-formed or floc-entrapped microplastics, secondary sedimentation may improve removal by preventing microplastic transfer from wastewater to the sludge stream. The basis for microbial reactions to the presence of microplastics may be laid by differences in microplastic abundance before and after biological treatment. Additionally, the variations in the number of microplastics before and after secondary sedimentation may provide light on the processes involved, such as retention in secondary effluent or transfer from floc formation and agglomeration to sludge treatment. Moreover, the redistribution of microplastics from the activated sludge back flow has not received much attention. Determining the mass balance of microplastics inside the biological treatment should reflect the actual elimination effectiveness of microplastics. Microplastics from the activated sludge back flow could cause an inaccurate assessment of the efficiency of biological treatment (Raju et al., 2020).

Future research should look at the mass balance of microplastics in biological treatment as well as the presence of microplastics in each unit operation and biological treatment process. Laboratory-scale research on the removal of microplastics using various biological treatments is still in its infancy. There are still questions regarding the removal methods of microplastics in various unit activities and processes or the corresponding impacts of microplastics on treatment performance. There are variations in reactions to the presence of microplastics due to the distinctive design of each unit operation and procedure in the biological treatment. Through agglomeration and floc formation, microplastics may indirectly interact with microorganisms in the suspended-growth biological treatment. However, laboratory-scale research rarely takes tank configuration and operational conditions of biological treatments into account. The length of the retention period, the tank's flow direction, and the oxygen availability may all have an impact on the amount of biofilm formation on microplastics or floc agglomeration with microplastics. Understanding the relationships between microorganism activity and microplastics can help explain the removal processes or pathways of microplastics. In contrast, microplastics in the attached-growth biological treatment tend to engage directly with microorganisms, acting as the growth medium for them. The types and forms of microplastic polymers may have an impact on the biofilm development. Rougher surfaces and aromatic groups on microplastics often make them better for microbe attachment. These characteristics initially controlled the adhesion of microorganisms to the surface, and the microbes released Extracellular Polymeric Substances (EPS) to aid in the attachment process. Ultimately, the surface of microplastics becomes home to a population of microorganisms that alter the material's surface characteristics. Depending on the degree of biofilm formation, the hydrophobicity, specific surface area, density, and functional groups of microplastics were altered. On the other hand, the microbial development on the surfaces of some microplastics may be inhibited by the leached additives from that particular polymer type. Different additives are included in different polymer kinds of microplastics.

7.10. INHIBITORY EFFECTS OF POLYMERS

The degrees of inhibitory effects on microorganisms should be assessed based on the amounts of leached additives from each type of polymer (Hermabessiere et al., 2017). Crucially, rather than using a dosage with known effects, the dosage of leached additives should be established based on the real concentrations of each polymer type of microplastic in primary effluent. These features should be taken into consideration when assessing the functions of microplastics on nitrification and denitrification in attached-growth biological treatment, as opposed to the inhibitory impacts of microplastics on treatment performance. Because the presence of microplastics does not hinder the action of microorganisms and essential enzymes, most studies have found minimal impacts of microplastics on nitrification and denitrification. Even at large dosages, the spherical or powder form of microplastics had no negative effects on microbial growth, abundance, or diversity. The intricacy of microplastics, taking into account their varied sizes, shapes, and polymer types, may impede the effectiveness of various biological treatments and, consequently, the removal efficiency. Different characteristics of microplastics may have an impact on the production of floc. In contrast to other polymer types of microplastics, which showed no discernible influence on either microorganism activity or treatment performance, PVC microplastics, for example, had a stronger negative impact on microorganisms and severely impeded treatment performance. Nevertheless, the proportion of PVC microplastics in primary effluent was minimal, and it was not thoroughly investigated how microorganisms actually reacted to the microplastics. When multiple types of microplastics are present, the microbial reaction may differ from that of a single kind of microplastic. Different forms and sizes of microplastics may have distinct mechanisms influencing the activity of microorganisms in addition to polymer type compositions. Larger or differently shaped microplastics cannot easily pass through the cell membrane than small, fibrous microplastics. Consequently, rather than promoting adhesion to microbial cells, they may have more inhibitory effects on microbes.

7.11. BIOLOGICAL TREATMENT

Before entering the biological treatment system, the initial sewage that enters the wastewater treatment plant will first go through primary treatment. Screening, the removal of sand and grease, and skimming are all part of the preliminary treatment process. After that, settleable contaminants are allowed to form sediment (primary sludge) as part of the primary treatment. While the majority of the microplastics in particle form are first adsorbed on the solids and then removed in the subsequent treatment stage, the separation of microfibers in sewage takes place during primary sedimentation. Fibrous microplastics are removed from the system during the primary sedimentation phase in a manner that is more substantial than that of synthetic particles. Smaller particles that are unable to penetrate primary sedimentation will form the influent composition of future treatment facilities, while bigger suspended debris is more readily retained by preliminary facilities and can be effectively removed by gravity-driven settling. The degree of microplastic clearance during primary treatment depends on a number of variables. One important aspect of sedimentation performance is the density of the microplastics that are visible in raw wastewater;

a higher density suggests a higher degree of settling. The amount of sewage that stays in the treatment facility is determined by its overflow speed, which also has an impact on the settling of microplastics. Additionally, the technologies used in the initial treatment systems have an impact on how microplastics are distributed throughout the main treatment system. To change the amount of microplastics imported into the primary settling tank, for instance, the grid and aerated grit chamber combination helps remove microplastics (Khadir et al., 2022).

7.12. MONITORING

Monitoring activities have shown that microplastics are present in the primary system's byproduct, or primary sludge, which is consistent with the removal of microplastics during the primary sedimentation process. Microplastics with distinct properties have been found across a broad range of values. Fibers make up a larger percentage of them than other types of polymers. Over 500 μm is the largest size for most microplastics. The primary chemical compositions of the microplastics found are acrylic, polyether sulfones, Polyamide (PA), Polypropylene (PP), and Polyethylene Terephthalate (PET). Furthermore, of these fiber polymers that are present in primary sludge, 72.2% had black as their predominant color. Despite the low proportion of plastics in this kind of sludge, they are multicolored, with blue being the most prevalent. The primary cause of the various colors observed in microplastics is that they come from distinct manufacturing and application methods. After passing through a first settling tank, the effluent next enters the secondary treatment system, which uses biological treatment to break down contaminants and aid in the sedimentation process. As a result, it is possible to eliminate the remaining microplastics further, however to a lesser degree than with the initial treatment. The biological system's extracellular polymers and microbial cells have a potent ability to adsorb microplastics. Microplastics may be present in sludge floc as a result of some bacteria (metazoa and protozoa) consuming the debris. Furthermore, some flocculants have the ability to collect microplastic pieces into flocs, which are then removed by precipitation.

Interestingly, in the subsequent secondary settling step, the unstable flocs of microplastics might separate after dispersing in the mixed liquid. In contrast to the primary treatment, the smaller microplastics (106–300 μm) in the influent can be removed more effectively because they are more easily adsorbed on the flocs and biofilms than the larger fragments (>300 μm) (Zhang et al., 2020). Additionally, during subsequent treatment, it was discovered that the effectiveness of microplastics is higher than that of fibers. This phenomenon is caused by the fact that most fibers with higher densities and larger volumes have already been extracted from the liquid in the primary treatment system; the remaining fibers, however, may be inert due to neutral buoyancy or other factors, making it challenging to remove them further. Remarkably, the presence of microplastics in waste-activated sludge may be determined by analyzing the microplastic contents in the influent and effluent of the secondary treatment system. The process of treating sewage may also have an impact on the retention of microplastics in waste-activated sludge. Particularly, varying secondary treatment system operating parameters result in a range of microplastic sizes in the effluent, suggesting that the particular microplastics formed in waste-activated sludge have distinct size distributions. Additionally, the duration of the microplastics' interaction with wastewater influences their removal in secondary treatment as well as their presence in waste-activated sludge. Longer contact times have been shown to have a role in the development of a biomass coating on the surface of microplastics. The microplastics may become denser as a result, which would facilitate their gravity separation.

Microplastic concentrations in waste-activated sludge have been found. Plastic particles that are less dense than water and challenging to separate during basic treatment are the principal components of waste-activated sludge. Therefore, waste-activated sludge is a collection of plastic particles smaller than 500 μm, and the majority of all microplastics are composed of plastics with diameters between 125 and 250 μm. There is also a comparison of the microplastic content in waste-activated sludge and primary sludge. Primary sludge is exposed to a higher concentration of microplastics from raw sewage than waste-activated sludge. This finding supports the notion that primary treatment has a more substantial role in the removal of microplastics. Furthermore, PET, PA, polyesters, and PP make up the majority of the microplastics found in waste-activated sludge, just like they do in primary sludge. It follows that these particles make up the majority of the microplastic content in the raw sewage (He et al., 2021). Waste-activated sludge also absorbed colorful plastics and fibrous microplastics, similar to primary sludge. Notably, the presence of microplastics in dewatered sludge is also impacted by different sludge dehydrating systems that operate on distinct principles. Sludge dewatering procedures can be classified as centrifugal dewatering or filtering dewatering depending on the mechanical forces utilized to extract the water from the sludge.

Some of the lighter microplastics will be moved from the sludge phase to the water phase during centrifugation, where they will be combined with raw sewage to be treated at the wastewater treatment system. Because of this, centrifugal dehydration produces dewatered sludge with light polymers than other widely used techniques. The geographical distribution of microplastics in dewatered sludge is primarily influenced by regional and climatic variations in forest cover, economic status, and urbanization; the temporal variability of microplastics is also influenced by variations in precipitation and temperature. Furthermore, rainfall and climate may have an impact on surface runoff, which in turn may have an impact on the presence of microplastics in water bodies. Diverse regions and climatic variables should be taken into consideration while choosing sludge treatment and disposal options.

7.13. METHANE AND PLASTICS

Additionally, sewage treatment system operating parameters can be optimized to regulate the concentrations of microplastics in sludge. Methane, a byproduct of the breakdown of sludge organics in an anaerobic digestion system, is frequently used as a critical metric to assess the efficacy of sludge treatment. Depending on the quantity, kind, and size of the microplastics, the effects on methane production from the anaerobic digestion of sludge have been investigated. Methane generation is inhibited by high concentrations of PE microplastics and PVC microplastics, respectively (Wang et al., 2023). The decreased methane potential and hydrolysis coefficient were found. An enhanced degree of inhibition is produced by higher concentrations of PE or PVC microplastics. Different quantities of PES microplastics show inhibition of methane generation, methane potential, and hydrolysis coefficient in an anaerobic digestion system, like PE and PVC microplastics. Nevertheless, there is no evidence of a relationship between the quantity of PES microplastics and the degree of methane production inhibition. However, it has been discovered that using 10 particles/g of polyamide 6 (PA_6) microplastics can effectively increase the amount of methane produced by 39.5% during sludge anaerobic digestion. Leaching of Caprolactam (CPL) from PA_6 microplastics is thought to be the primary cause of this occurrence, as it may stimulate the activity of important enzymes involved in acidification and methanogenesis. A marginal increase in methane production from sludge anaerobic digestion at low PVC microplastic concentrations may have been caused by improved solubilization. Therefore, the amount and kind of microplastics present determine how they affect the production of methane from sludge anaerobic digestion. Another important consideration for microplastics' effects on the anaerobic digestion of sludge that produces methane is their size. Smaller microplastics have a higher probability of penetrating a cytomembrane and seriously harming microorganisms. Furthermore, some nanoscale toxicity investigations have been carried out. Exposure to nanoplastics suppresses the maximum daily methane yield and methane production; the inhibitory effect increased with increasing nanoplastic concentrations. In addition to anaerobic digestion, anaerobic fermentation is another popular technique for treating sludge. This process produces high-value products known as short-chain fatty acids or SCFAs, such as acetate, propionate, butyrate, and valerate, which are proven to be the right carbon source for wastewater treatment.

Many researchers are currently concentrating on the investigation of the effects of microplastics on the anaerobic sludge fermentation process' ability to produce SCFAs. The generation of SCFAs is significantly impacted by various forms of microplastics (Fournier et al., 2023). For instance, compared to the control group without microplastics, the synthesis of SCFAs is increased at all evaluated doses of PA6 microplastics, mostly because the enzyme activities involved in acidification. On the other hand, the maximum SCFA generation is inhibited when mixed microplastics, such as PE, PET, PS, and PP, are added to the sludge anaerobic fermenter. The inclusion of microplastics obviously suppresses both acidification and hydrolysis, which lowers performance. Likewise, it has been demonstrated that PS microplastics inhibit the synthesis of SCFAs by interfering with the hydrolysis and acidification processes. Particularly, 10 μg/L of PS microplastics shows a stronger inhibitory effect than 50, 100, and 1000 μg/L of PS microplastics. This is probably because there is less aggregation and more attachment at lower PS microplastic concentrations. However, PS nanoplastics increase the production of SCFAs through an increase in acetate and propionate, they have a detrimental effect on the system's ability to form hydrogen.

Another well-known byproduct of the anaerobic treatment of sludge is hydrogen. We are still in the early stages of researching the possible impact of microplastics on hydrogen production in sludge treatment. The presence of PET microplastics inhibits the formation of hydrogen from WAS (waste activated sludge) alkaline anaerobic fermentation, and this degradation is increased with an increase in PET microplastic concentration. Additional research revealed that the inhibitory effects on the hydrolysis, acidogenesis, and acetogenesis steps of WAS alkaline anaerobic fermentation may be primarily caused by the toxic Di-n-Butyl Phthalate (DBP) leached from PET microplastics and excess oxidative stress induced by PET microplastics. Additionally, researchers investigated how PS nanoplastics affect the hydrogen bioenergy that is recovered from sludge using hydrogen fermentation of the acetic acid type. Based on the findings, 0.2 g/L PS nanoplastics reduce the protective effect of EPS by degrading its surface charge, composition, and structure. This leads to a reduction in cumulative hydrogen production at 53.9% of control. PS nanoplastic exposure increases cell permeability and exacerbates the antioxidant system's imbalance, lowering the cell's resistance to outside stimuli. Enzyme activity plays a role in the complex metabolic process of anaerobic sludge treatment, which is dependent on the involvement of several enzymes. Common hydrolases such as protease, cellulase, α-amylase, and α-glucosidase are essential for the transformation of proteins and polysaccharides into amino acids and monosaccharides. The two main enzymes involved in the bioprocess of acidogenesis are Butyrate Kinase (BK) and Acetate Kinase (AK).

While AK takes part in the bioconversion of acetyl-CoA to acetic acids, BK catalyzes the conversion of amino acids to SCFAs. Furthermore, coenzyme F_{420} is necessary for the methanogenesis stage. It has been found that the activities of protease, AK, and F_{420} can be impacted by the presence of PVC and polyether sulfone microplastics. F_{420} was the most sensitive of all. Nonetheless, moderate concentrations of PA6 microplastics have been found to positively impact the activity of important enzymes such as α-glucosidase, protease, AK, BK, and F_{420}. F_{420} activity has improved twice as much as it would have in the absence of PA6 microplastics. In this case, the CPL leached from PA6 microplastics is essential. When CPL and enzyme molecules combine, the active site of the enzyme is changed, increasing the catalytic activity of the enzyme. Regarding nanoplastics, it was discovered that when 0.2 g/L PS nanoplastics are present in the system of hydrogen recovery by thermophilic bacteria, (Zhang et al., 2022) the activity of α-amylase drops to 61.8% from that of the system without nanoplastics, preventing sludge hydrolysis and hydrogen production. The enzymes (antioxidant) can be impacted by microplastic exposure in addition to the enzymes governing each bioprocess engaged in anaerobic digestion. Microorganisms typically defend against mild oxidative stress through a specialized antioxidant mechanism. Enzymes that scavenge certain ROS include Catalase (CAT) and Superoxide Dismutase (SOD). However, due to the small number of bacterial enzyme molecules, this neutralization is unable to protect against ROS at large concentrations. Exposure to PS nanoplastics inhibited the activities of SOD and CAT, which reduced the ability of microorganisms to withstand the severe oxidative stress that PS nanoplastics cause.

7.14. MICROPLASTICS ALTER THE BACTERIAL ABUNDANCE

Microplastics alter the relative abundance within the microbial community, albeit with minimal effects on the diversity and structure of the population. The tolerance that different microorganisms have for microplastics varies. The predominant phylum in methane-producing anaerobic reactors is *Euryarchaeota*, a well-known archaeal phylum that contains methanogens; its abundance decreases during exposure to microplastics. Aceticlastic *Methanosaeta* sp. is a type of methanogen that produces methane by using acetate; its relative abundance decreases in the presence of microplastics, which is consistent with a fall in the efficiency of methane production from sludge. *Methanoregula* sp. and *Methanomas siliicoccus* sp. are anaerobic methanogens that use methanol or methylamine, H2, CO2, and other resources to produce methane. When metal-doped nanoplastics are added, their relative abundances decrease, suppressing methane production during anaerobic digestion. The main bacterial phyla in sludge anaerobic treatment systems include *Euryarchaeota*, *Proteobacteria*, *Bacteroidetes*, *Firmicutes*, and *Actinobacteria*. Several anaerobes in these phyla are capable of converting organic molecules into short-chain fatty acids (SCFAs). When exposed to microplastics, most of these bacteria often exhibit a drop in abundance. Furthermore, exposure to microplastics has been shown to decrease the numbers of important bacteria involved in the decomposition of organic materials, including acidogens and hydrolytic microorganisms. Microplastics can alter the relative abundances of significant bacteria and physically damage microorganisms to varying degrees. Microplastics and microbial cells may interact to impair extracellular polymeric substances (EPS) defenses or potentially damage the integrity of cell walls and membranes, thereby reducing cell viability. Sludge from by-products is produced in enormous quantities by wastewater treatment plants.

7.15. TREATING MICROPLASTICS IN WASTEWATER

One common method for appropriately disposing of treated wastewater treatment plant sludge is land application and landfilling. As a result, there is a chance that the sludge will carry microplastics into the ground and landfills. Applying treated sludge to the soil to increase fertility is known as land application of sludge. Large amounts of organic matter and mineral components like potassium, phosphorus, and nitrogen are present in the sludge, which can help the soil meet its nutrient needs and even improve it to some degree, increasing soil fertility. This approach for reusing sludge has been widely applied in soil and has been shown to be beneficial to the circular economy. Sludge application is promoted by the agricultural industry in Spain, where 65% of the sludge is recycled through the soil (Mosquera-Losada et al., 2017).

Adding sludge to soil creates a new pathway for microplastics to enter, and poor sludge disposal can worsen the buildup of microplastics in the soil. Sludge is utilized in agriculture to the tune of 45% in China. The overall amount of microplastics accumulated in the soil grows with time as a result of the long-term application of sludge products containing microplastics, and this amount is positively connected with the frequency of sludge application. In Denmark, farms that reuse their sewage have twice as many microplastics as farms that don't apply sludge. Since microplastics degrade slowly, it is possible to find them in soil up to 15 years after sludge application.

Microplastics carry a number of possible risks when they get into the soil. For instance, they may have an effect on the soil's bulk density, water-holding capacity, and other physical and chemical characteristics. The soil biota will be impacted by the accumulation of microplastics in plants and earthworms. Furthermore, the majority of the microplastics in the sludge have rough surfaces. The adsorption potential of microplastics in sludge for metal contaminants is ten times greater than that of fresh microplastics. When sludge is applied to land, the environmental risk posed by the microplastics within it could be greater than that of newly formed microplastics. This is a problem that needs a lot more work. More significantly, the majority of microplastics that enter the soil through agricultural soil amendments will migrate from the soil to the aquatic environment, where they will increase environmental risks. Less than half of these microplastics will stay in the soil. The most common method of disposing of sludge is through landfills. According to estimates, 79% of plastic produced is accounted for by the microplastics that are dumped in landfills from sludge. As a result, a landfill will gather a lot of microplastics because it is a significant location for the discharge of sludge. Only 2% of the sludge is ultimately landfilled in Australia, compared to 34.5% in China. 22% of the sludge is disposed of in landfills, which means that up to 464 trillion microplastics are dumped there, according to data from the US Environmental Protection Agency. The environment will be negatively impacted by microplastics that are transported to the landfill by sludge because they will blow out of the dump and onto the surrounding region. The breakdown of waste produces different types of leachates when the landfill is in operation. Pollutants such as organic and heavy metals are abundant in the landfill leachate. The ability of newly discovered polluting microplastics to transport other pollutants has been discovered in recent years, which exacerbates the harm that leachate discharge causes to the environment. There will be more potential risks if they are not handled correctly.

One of the important sludge treatment and recycling processes is composting. There is growing evidence that microbes play a significant role in the breakdown of plastics. Certain strains of Arthrobacter and Bacillus that were isolated from the plastic surface of marine waste have been found to degrade polyethylene plastic. Simultaneously, researchers were able to successfully isolate bacteria that break down polyethylene from the biofilm that formed on the sediment microplastics' surface. Plastics can biodegrade through the composting process when compared to the natural environment. Microplastics are mostly in the form of flakes following sludge composting. This is because aerobic composting at a high temperature of 70°C breaks down microplastics. Microplastics in sludge are affected by both conventional composting and ultra-high temperature composting (Sun et al., 2023). After 45 days of treatment, the microplastic content using full-size ultra-high temperature composting (70°C) dropped by 43.70%, ten times more than that of conventional composting. After 56 days of degradation in the ultra-high temperature composting group, the analysis of the composting bacterial population shows that the hyperthermophilic mixed bacteria cause 7.3% weight loss of PS microplastics, which is six times more than that of the conventional high-temperature (40°C) group. The biodegradation of plastics is accelerated by the breakage of C–C bonds at high temperatures. Sludge is frequently reduced via incineration, which turns it into a useful and affordable mineral product. Plastics are entirely eliminated, particularly after burning, and are no longer able to pollute the environment. However, the incineration process uses a lot of energy and produces hazardous and toxic gases like dioxins due to the high-water content and the enormous amount of microplastics in the sludge.

MPs are being found in a variety of settings due to their growing use and applicability. According to a survey by the US National Research Council, effluents are believed to be responsible for 70–80% of the 580 Mt of waste that has been dumped into the oceans. Furthermore, it was mentioned that regions with high concentrations of MPs and Phthalic Acid Esters (PAEs) were typically those that were close to anthropogenic activity. All samples obtained from an aquatic system in the Indian subcontinent have been shown to contain microplastics, ranging from 2–64 particles/L in water samples to 15–632 particles/kg/dw in sediment samples. Many countries, including the USA, China, and Korea, have observed this behavior to varying degrees. Additionally, scientists are starting to focus more and more on the substances connected to MPs. MPs from large cities had substantially greater amounts of PCBs, PAHs, and HCHs than MPs from suburbs or small cities. Specifically, MPs from beaches and the coast of Sao Paulo State were found to have over 2000 ng/g of Polychlorinated Biphenyls (PCBs). Additionally, environmental scientists concentrate on the factors that influence MPs and chemicals. The high octanol partition coefficient of MPs indicates that hydrophobic organic contaminants, like biphenyls and Polycyclic Aromatic Hydrocarbons (PAHs), are more readily adsorbed on them (Wang et al., 2020).

The primary MPs found in the environment are Polystyrene (PS), Polypropylene (PP), Polyethylene (PE), and Polyvinyl Chloride (PVC). Because of their differing polarity and functional groups, PE often has higher sorption affinities for organic compounds than PP and PS. The sorption behavior between MPs and organic pollutants was also influenced by salinity, pH, and other environmental factors. While pH is another significant environmental element that may have an impact on adsorption by altering the structure of MPs and their electrostatic interactions, salinity is advantageous to the adsorption of chemicals by MPs and its influence was related to chemicals' characteristics. There is minimal research on the interactions between chemicals in MPs and their fragments, the release of chemical additives for MPs, and their environmental behaviors. The adsorption kinetics of MPs for chemicals are mostly related to the surrounding environment and influencing factors. The interaction between pollutants and MPs in an aqueous environment depends on the adsorption process. Thus, a summary of the studies on MPs' adsorption behavior toward environmental pollutants is crucial. To demonstrate the adsorption behavior of MPs towards environmental toxins, representative contaminants like organic compounds, heavy metals, and antibiotics were chosen for research studies. Antibiotics that are present in different settings and are utilized as pharmaceutical and personal care items have drawn a lot of attention due to their ecotoxicity and environmental degradation. Tetracyclines (TC), Macrolides (MCL), Fluoroquinolones (FQL), and Sulfonamides (SDZ) are the antibiotic classes that have been found in aquatic habitats. Because MPs and antibiotics are so widely used together, their combination is unavoidable and harmful to aquatic life.

The function of MPs as antibiotic adsorbents has been the focus of numerous investigations. Hydrophobic, hydrogen-bonding, and electrostatic interactions are the main adsorption mechanisms of MPs for antibiotics. Furthermore, several processes, including halogen bonding, CH-π interactions, cation-π interactions, and negative charge-assisted hydrogen bonds, are also used to explain the antibiotics' adsorption. Antibiotic sorption on MPs varies depending on the kind of MP and antibiotic. MPs can adsorb a variety of antibiotics, such as Ciprofloxacin (CIP), Trimethoprim (TMP), Amoxicillin (AMX), etc. Some researchers investigated the adsorption capability of different types of MPs for f different antibiotics in both freshwater and seawater and found that the amount of MPs adsorbed reduced as the antibiotics became more hydrophobic. PS had better adsorption properties than PE for CIP, TMP, and SDZ. This is most likely since PS and PE had π-π interactions at the aromatic surface. Similarly, some researchers looked into the Sulfamethoxazole (SMX) adsorption of MPs; they found that PA had the maximum SMX adsorption capability (Yu et al., 2023).

7.16. MECHANISMS OF MICROPLASTIC DAMAGE

Additional research has revealed that the adsorption of specific pollutants, such as heavy metals and dissolved organic matter, might increase the affinity of antibiotics for MPs, improving the interaction between MPs and antibiotics.

Heavy metals are also adsorbed on MPs,. Environmental contamination by heavy metals is thought to pose a serious global concern. It was estimated that heavy metal pollution endangered over 40% of the lakes and rivers. Excessive concentrations of heavy metals can have a variety of negative impacts, including harming ecosystems, altering an organism's growth and metabolism, and even posing a risk to human health and animals farther down the food chain. Several investigations revealed that MPs could absorb heavy metals from the surrounding environment and then transfer them to the aquatic biota. Understanding the behavior and mechanism of such adsorption of heavy metals is therefore crucial.

Programs for monitoring the environment revealed that heavy metals were present and accumulated on MPs. Because MPs had a larger surface area and higher polarity than the surrounding aquatic environment, their surface displayed a greater concentration of heavy metals. The most often found heavy metals on MPs were Zn, Fe, Pb, and Cu, which were 2–13,400 times more concentrated than those found in the aquatic environment. It was discovered that the amounts of heavy metals in MPs differed by area, maybe based on the source of the heavy metals. It is noteworthy that Ti showed the highest MP adsorption concentration of all the metals in the Beijiang River (China) and Lake Garda (Italy).

Furthermore, Cr was one of the most widely used heavy metals on MPs in India, but not in Thailand or China. All of these investigations demonstrate that MPs can absorb ambient heavy metals and that the adsorption behavior changes depending on the MP. Laboratory studies were conducted to investigate how MPs absorb heavy metals. Because MPs are so small and common in aquatic settings, it is evident from the data that there is a significant cause for concern. MPs can absorb heavy metals from their surroundings onto their surface. This implies that MPs carry heavy metals and are essential to the movement of heavy metals through the environment.

Several investigations have shown that MPs have a considerable capacity for adsorption of various organic pollutants. Pharmaceuticals, personal hygiene items, endocrine disruptors, pesticides, and typical industrial organic wastes like phenolics, halogens, and aromatics are examples of environmental organic pollutants.

The composition of organic contaminants in water bodies is complex, variable, and generally dangerous. Therefore, in order to comprehend the ecological risk of their combination, it is important to discuss MPs' adsorption behavior for organic contaminants. Because of adsorption processes, a number of investigations have shown that greater concentrations of organic contaminants than are environmentally important were recorded on MPs surface. Similar to heavy metals and antibiotics, the adsorption capacity of MPs differed depending on the kind of MP and the organic contaminants. Antibiotic-resistant bacteria were reported to have infected over 3.1 million patients from 96 different regions of the world in 2020, and this figure is predicted to rise in the next few years. This issue is made worse in particular by gene exchange between ambient microorganisms and Antibiotic-Resistant Bacteria (ARB). Antibiotic Resistance Genes (ARGs), the genetic carrier of antibiotic resistance, are regarded as emergent pollutants and have been found in freshwater, wastewater, and marine ecosystems, among other aquatic systems. MPs were discovered to have the ability to enhance ARB and ARGs on their surface (Syranidou and Kalogerakis, 2022) which would increase their propagation and pose a serious threat to the aquatic environment. Therefore, it is necessary to describe the adsorption/enrichment performance and mechanism of ARGs and ARB on MPs.

The MP surface becomes colonized by microorganisms, which then produces a biofilm known as a plastisphere. Concern over the plastisphere in aquatic ecosystems has grown recently since it has been discovered that the microbiota within the plastisphere differs from the surrounding environment in terms of composition and diversity, and is even selective for ARBs and ARGs. Several investigations have shown that MPs have a great affinity for ARGs. Additionally, MPs affect the distribution and makeup of the microbial community inside a system, particularly concerning ARB. There were more ARGs on MPs than in the surrounding area.

MPs enriched ARGs and ARB on their surface, as evidenced by the high correlation between ARG profiles and microbial diversity. Furthermore, researchers noted that more unusual ARB, such as *Pseudomonas syringae*, *Pseudomonas mendocina*, and *Pseudomonas monteilii*, were only found in the microbiota of the plastisphere and not in biofilms covering other natural materials (leaves), suggesting that MPs functioned as a unique niche for the selective enrichment of ARB rather than other materials. Furthermore, it was shown that MPs improved the development and metabolism of the microbiota on plastispheres, implying that the enriched ARB on plastispheres proliferated more quickly than in the surrounding environment.

The quantities of ARB and ARGs in an aquatic system are raised by the enrichment of ARB and the increased microbial turnover in the plastisphere, which poses a risk to the aquatic environment and human health. Additionally, gene exchange, also known as Horizontal Gene Transfer (HGT), can promote the dissemination of ARGs. Because this surface biofilm creates the perfect environment for cell-to-cell communication and the transmission of ARGs between ARB and ambient bacteria, MPs may have an impact on the gene exchange of aquatic ecosystems. Microplastic-associated bacteria showed a higher frequency of plasmid transfer than free or naturally aggregated bacteria. Several Mobile Genetic Elements (MGEs) were also discovered to be present on the surface of MPs. Their abundances were greater on the plastic surface than in the nearby water. When MPs get together, they may attract more ARB to their plastisphere and create a unique microbiome that is more abundant in ARB than the surrounding air. Additionally, MPs offer an environment that is conducive to the growth, proliferation, and exchange of absorbed ARBs and ARGs, which facilitates the spread of both ARBs and ARGs in the aquatic ecosystem.

Chapter 8

HUMAN HEALTH AND PLASTICS

Contents

Introduction 168
8.1. Toxicity to the Lungs 168
8.2. Skin Damage 169
8.3. Sensitivity 170
8.4. Carcinogens 171
8.5. Endocrine Disruptors 172
8.6. ECD Concerns 173
8.7. Proper Evaluation 175
8.8. Colorants-Concerns 176
Conclusion 178

INTRODUCTION

Currently, the number of synthetic chemicals in use exceeds 70,000. The majority of these substances are classified as toxic or dangerous to human health, but only 7% of them have ever undergone sufficient testing to ascertain their impact on both humans and other living things (Krewski et al., 2010). We come into contact with, use, inhale, and swallow a variety of these chemicals directly or indirectly, both within our homes and outdoors where we spend the majority of our time. Each of these chemicals has the potential to have varying effects on human health. Although we humans are aware of these impacts and the risks they pose, the facts are still disregarded in certain situations. These consequences might be direct or indirect. Australia and the United States are the most industrialized countries with the highest emissions per capita. This served as a supplement to the Kyoto Protocol, which was ratified by 145 countries and went into force in February 2006, requiring industrialized nations to reduce their emissions. It's interesting to note that while Australia and the United States are not yet parties to the Kyoto Protocol, Japan is a supporter with legally binding commitments, and China, India, and South Korea are likewise supporters with non-binding commitments, China having the fastest-growing economy among the top six. Fogging of automobile windows on the inside is an issue, particularly in newer cars. This is also caused by Volatile Organic Compounds (VOCs) released from Polyurethane (PU) foam in seats and Polyvinyl Chloride (PVC) seat covers, among other sources. New rubber tires typically have a bad odor. The unpleasant odors and fogging, which indicate the emission of different VOCs from these processed items, are likely also indicators of toxicants. These can be identified and measured at very low concentration levels of ppm (parts per million) or even ppb (parts per billion), which are well documented, using standard analytical methods. However, because some substances with low acute toxicity might have carcinogenic or endocrine effects even at very low doses that show no indication of acute toxicity at all, LD50 values may not always adequately reflect the complete spectrum of toxicity. The FDA has set the safe level for humans at 1 ppm if a test animal has a threshold of 100 ppm, assuming that humans are ten times as sensitive to the material as animals and that the weaker segments of the population are about ten times as sensitive as the healthy population.

8.1. TOXICITY TO THE LUNGS

When poisonous gases or Volatile Organic Compounds (VOC) vapors are present in the atmosphere, people inhale them. These substances are known as gaseous toxins. The size of the molecules in gas and vapor is different, measuring 0.0005 m for gas and 0.005 m for a vapor. Hazardous smoke or fumes may also be present in the air, with smoke particles from 0.01–1 μm, and fume particles from 0.01–1m. The target organism absorbs toxins through inhalation, initially in the upper tract of the nose and subsequently in the lungs. For this method, the nose serves as a chemical scrubber. As a result, gases that are highly soluble in water, such as hydrogen chloride and ammonia, can dissolve in the mucous coating (wetness) and cause irritation. Chlorine and other substances with intermediate water solubilities can irritate the respiratory tract at any point, while nitrogen dioxide and phosgene, two insoluble substances, can enter blood vessels. The most hazardous point of entry is inhalation since the lung membrane is ineffective as a barrier. As soon as gas molecules enter the lower tract, they can diffuse through the lung membrane in less than 75 seconds, enter the capillary network, and dissolve into the blood to be transported throughout the body. The concentration, water solubility, depth of breathing, and blood circulation rate all affect how quickly harmful compounds are absorbed. Owing to the blood vessels and respiratory system fragility, an inflammatory reaction occurs, leading to numerous lesions being identified as bronchiolitis, and so forth. Depending on the response, the effects can be acute, subacute, or chronic. When it comes to airborne hazardous particulates, larger particles (get trapped in the upper airways or upper respiratory tract, whereas smaller particles can enter the blood vessels and lower airways or lower respiratory tract, respectively.

8.2. SKIN DAMAGE

Because of the skin's porous structure, some harmful compounds can be absorbed by it and enter the bloodstream. This is the most typical exposure path. Skin structure can vary in permeability as well, from the delicate (scrotum) to the rough (soles). Skin serves as an excellent barrier to prevent the entry of chemicals, or inorganic substances, but cuts can hasten the absorption process. The ease of absorption of organic chemicals varies depending on the environment; in fact, some organic molecules, like dimethyl sulfoxide (DMSO), can facilitate the absorption of other organic chemicals through the skin. It has been demonstrated that the eyes are also very good routes for the body to absorb toxins. Anywhere along the gastrointestinal tract, chemicals that are consumed have the potential to enter the bloodstream. The substance will almost certainly be removed from the body in some way if it cannot be absorbed by the body. One crucial element in evaluating a chemical's harmful effects is what happens to it once it is introduced to an organism. It includes any scenario in which the chemical and other chemicals currently present in the organism could interact or have joint/cumulative effects. Regarding the VOC, the blood-to-gas-phase partition coefficient values, tissue reactivity, and water solubility of the toxins are all significant. The body stores many poisonous/toxic compounds, primarily in fat or bone, which can circulate throughout the organism for extended periods of time. Toxic effects and their severity are typically determined by species differences/variations in metabolism. The development of selective pesticides, which can kill specific pests while being relatively harmless to the user and other animals likely to come into contact with them, is made possible by the fact that different species have different metabolisms and physiological conditions, which can lead to some chemicals being harmless to certain vertebrates but toxic or highly toxic to others. One example of such a compound is norbormide, a heterocyclic nitrogen compound that is extremely poisonous to rats but negligibly low toxicity to humans and domestic animals (Zakharychev et al., 2020).

8.3. SENSITIVITY

The target function or organ remains constant, notwithstanding individual differences in chemical sensitivity. Additionally, keep in mind that results from animal studies that provide reference toxicities are not necessarily definitive when applied to humans. The results observed in animals may not always indicate that humans should experience the same outcomes (toxicogenomic approach). A single, brief exposure to an acutely hazardous chemical can have negative effects. In general, though, in a small number of instances, it might also have an impact long after the initial exposure. Since both types of toxicity have dose-related effects, the larger the dose, the bigger the damage. It should be mentioned that not all compounds that are poisonous in the short term will cause long-term damage, and not all substances that are toxic in the long run will cause problems at a single dose. The substances known as carcinogens, sometimes known as carcinogenic, are most frequently linked to long-term damage. It is necessary to specify short-term (less than 24 hours) and long-term (more than 24 hours) time spans. Thus, exposures that occur repeatedly or continuously and last for up to 24 hours are classified as acute hazardous exposures. People who suffer from specific medical conditions (liver or lung disorders) are more likely to be impacted by exposure to poisonous substances and to respond more severely afterward. Anytime a chemical is absorbed, it undergoes some sort of change or metabolism that results in detoxification, that is, either by forming less hazardous metabolites or by breaking down into components that can be integrated or expelled. The absorbed chemical and its metabolites may accumulate in a particular tissue, such as the liver, kidney, adipose tissue, or other tissue, or they may be expelled, stored, or moved within the organism and reach locations where destructive consequences may be generated. Rapid excretion often has a modest toxicity effect. A slower rate of excretion may result in more detrimental long-term effects. As long as there is no residue left, subsequent intakes of the same doses are also eliminated in the event that the previous dose is completely voided. If dosages are not repeated frequently enough, the residue from the second dose may combine with the first and become poisonous.

8.4. CARCINOGENS

A carcinogen is an external substance or agent that can induce cells or tissues to go through several uncontrolled growth stages, resulting in malignant neoplasms. Cancer is one of the three main causes of death in the world. A chemical is deemed carcinogenic if it has been assessed by the World Health Organization's International Agency for Research on Cancer (IARC) and determined to be a carcinogen or a potential carcinogen; if it is included in the National Toxicology Program's (NTP) Annual Report on Carcinogens; or if the Occupational Safety and Health Administration/OSHA has recognized it as a carcinogen. The majority of cancer causes (80–90%) are attributed to a variety of known or probable carcinogens, or outside environmental variables, such as cigarette smoke. Numerous internal variables, including irradiation (radioactive, ultraviolet, and radiofrequency), certain viruses (polyoma), and weakened immune systems from old age, poor health, inheritance, etc., can also cause cancer. Since carcinogens are regarded as a unique class of chronic poisons/toxins, they are the focus of a specific area of toxicology. Regarding the cancer issue, not everyone reacts to external and internal variables in the same manner due to the variety of elements at play. Carcinogens can behave similarly to other harmful compounds, with comparable dose-response relationships but some notable variances. Examples of these substances are organic and inorganic chemicals with diverse biological effects. A carcinoma is any type of cancer that starts in the epithelial cells. Carcinomas are defined as malignant since they encroach on nearby tissues and organs and have the potential to migrate to distant locations and lymph nodes (metastasis). A precancerous condition or Carcinoma *In Situ* (CIS) occurs when there are cytological indications of cancer but no histological proof of penetration through the epithelial basement membrane. A WHO report states that about thirty percent of carcinogenic compounds come from smoking and that thirty percent come from chemicals related to food and drink (Gold, 1990). There is a lengthy active list of chemicals that are known to cause cancer and are categorized as likely or probable. The list is updated on a regular basis and includes both expected and unexpected chemicals that cause cancer, like some tranquilizers and antibiotics, antipyretics, analgesics, etc. that may cause liver dysfunction and eventually cause liver cancer. The risk numbers for substances are typically allocated, such as R_{45} (which may cause cancer generally) and R_{49} (which may cause cancer by inhalation). Additionally, several anticancer medications are said to be able to cause carcinomatosis on their own. Numerous agricultural chemicals have been shown to naturally cause cancer. The connection between chemicals and the disease was established in the late eighteenth century when cancer was first identified in individuals who worked as sweepers in the United Kingdom. Subsequent research revealed a connection between polynuclear aromatic hydrocarbon structure and its carcinogenic potential.

8.5. ENDOCRINE DISRUPTORS

The endocrine system, also known as the hormonal system, is a network of glands and receptors that regulates many important bodily processes, including growth, development, and maturation, as well as the operation of different organs in vertebrate bodies. The endocrine system secretes hormones that serve as the primary means of communication and regulation between the neurological system and every bodily function. Endocrine Disruptors (ECD) are a class of compounds that can interfere with the endocrine system's regular operation. Following its absorption by the body, an ECD typically mimics or blocks hormones, disrupting the body's regular activities. The majority of ECDs are synthetic chemicals that, at very low concentrations, induce hormone-related disorders, primarily linked to reproduction and dysfunction. In humans and wildlife, exposure to any of these compounds in the environment typically causes disturbance of thyroid hormones, androgens, estrogens, and other endocrine functions.

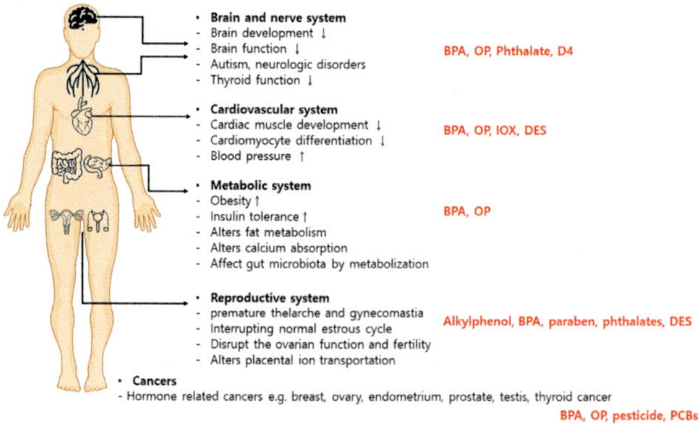

Figure 8.1. *Endocrine disrupting chemicals.*

Source: Ahn C, Jeung E-B. Endocrine-Disrupting Chemicals and Disease End-points. International Journal of Molecular Sciences. 2023; 24(6):5342. https://doi.org/10.3390/ijms24065342, Creative Commons License.

The production of synthetic estrogens can occur by burning, industrial production, or as byproducts of such operations. Laboratory investigations, such as those measuring a chemical's capacity to accelerate the growth of cultures of breast cancer cells, have identified known synthetic estrogens. There is still more to learn and research to be done regarding the role of ECD, as its mechanisms are not well understood and its specific endpoints or effects are not fully defined. While adult exposure to ECD can be reversible, the disruptive consequences in this situation are demonstrated to be permanent and irreversible, making the effects of ECD on developing organisms of particular importance. Many different organic compounds, the majority of which are synthetic and include additives and plasticizers like phthalates, bisphenol A, and nonylphenols, which are all used as additives in plastics, are thought to be possible ECD agents. In addition to many industrial and consumer items including liquid soaps, shampoos, conditioners, and hair colors that include Alkylphenol Ethoxylates (APE), which have been replaced with the more costly but safer alcohol ethoxylates, the list of ECD also includes pesticides like the prohibited DDT. Moreover, dioxins, certain preservatives, and metal ions have been linked to human endocrine disruption based on scant animal research. These substances include Polychlorinated Biphenyls (PCBs).

8.6. ECD CONCERNS

The main source of concern for ECD at the moment is synthetic estrogens; however, there are some natural estrogens, or phytoestrogens, that are safer (Autrup et al., 2020). They are found in many different types of plants in nature, and because they are safely metabolized or degradable, they do not bioaccumulate in living things. It is thought that the endocrine system can be disrupted. A natural hormone, such as a growth hormone that causes an increase in muscle mass, can be mimicked by some ECD, tricking the body into overreacting to the stimulus or responding at the wrong times, such as when insulin is not needed. Certain ECDs can prevent a hormone's effects on specific receptors such as growth hormones, which are necessary for appropriate development. Some ECDs directly stimulate or inhibit the endocrine system, which can result in either an overactive or underactive thyroid, among other hormonal overproduction or underproduction issues. These effects are all undesirable. Nevertheless, ECD can also be beneficial in humans. For example, birth control pills are commonly used medications that are known to have ECD. The environmental and medical sciences are particularly interested in ECDs; however, while some data regarding the impact of pesticides on ECDs is accessible, most chemicals have unknown ECD-related risks, and there are several scientific ambiguities. There are numerous instances of how ECD can have an impact. Diethylstilbestrol (DES), a synthetic estrogen that was once a powerful medication, is typically cited as one such instance. Although doctors provided DES to approximately five million pregnant women in the early 1970s to prevent abortion or to encourage fetal growth, the drug is currently prohibited. DES actually has an impact on the development of the reproductive system and can lead to cancer. In light of these instances, it is currently necessary for any new application to go through the appropriate ECD screening procedures. Workers producing synthetic leather were exposed to Dimethylformamide (DMFA) for one to five years, which was found to have a negative impact on conception, pregnancy, and delivery, raising the risk of birth problems. Studies demonstrated that exposure to styrene monomer is linked to several disorders, including hypermenorrhea, chronic sickness with secondary amenorrhea, and nulliparity with less blood clots.

Prior to being marketed, synthetic chemicals are typically tested, and the results are utilized in the second level of risk assessment, which is followed by risk management. The consequences of ECD on reproduction and disorders that are linked to it, like cancer, are taken into consideration by current EU regulation, but it does not address any other endocrine disruptions. PCBs have remarkable thermal and chemical stabilities, they were mostly utilized as plasticizers from World War I till now in electrical equipment such as transformers and capacitors, and hydraulic fluids. Large amounts of PCB are thought to still be present in some outdated transformer and capacitor systems, even though their production was discontinued in the USA in 1977 (Erickson and Kaley, 2011). Undoubtedly, PCBs can still be found in some electrical equipment in the USA. They are also commonly discovered in contaminated sediments and hazardous waste sites across the globe. PCB was present in sealants made of polysulfide polymers that were used in buildings 20–40 years ago. It has been demonstrated that PCB is still present in some Swedish homes at disturbing levels. Research has demonstrated that offspring that are exposed to low concentrations of perchlorate-containing fish while their mothers are still pregnant experience low IQ, impaired reading comprehension, and memory issues. There are about seventy five dioxins and 135 related furans in the dioxin family, which is the most dangerous group of chlorinated organic chemicals that we are currently aware of. These are ECDs for humans and can cause cancer even at extremely low exposure levels because they are easily soluble in body fat. The toxicity of a molecule is greatly influenced by the number and location of chlorine atoms, with seventeen dioxins being classified as extremely dangerous. These include dibenzofurans and polychlorinated dioxins, which are byproducts of burning chlorinated hydrocarbons (PCB, PVC, and pentachlorophenol), burning municipal and medical waste, and natural disasters. These harmful compounds are readily absorbed into the soil, where they can bioaccumulate in fish and other wildlife. In fact, the levels of dioxin in fish are typically 100,000 times higher than those in the surrounding environment. Therefore, the food chain is the easiest way that humans are exposed to dioxins. When PCBs are present, combustion and incineration processes create and release PCBs with four or more chlorines and one or no replacement in the ortho position, which are thought to have dioxin-like toxicity. It should be noted that planar molecules with considerable toxicity are produced by para or meta substitutions. Persistent Organic Pollutants (POP) are specific organic compounds, primarily polycyclic aromatic hydrocarbons, that can cause cancer directly or through metabolism. They can also harm human health at very low concentrations by being inhaled, ingested with food, or absorbed depending on temperature. Two well-known POPs include dioxins and PCBs. The Stockholm Convention applies to all POP and requires signatory nations to take appropriate action to minimize, if removal is not feasible, or completely eradicate all sources of dioxins.

A complicated collection of specific chemical compounds and minerals, additives make about 15–25% of the weight of all marketed plastic goods (Turner and Filella, 2021). In order to improve and create a variety of distinctive performance characteristics in plastics and shield them from the impacts of heat, and environmental factors, additives are essential. There might be some compounds in the system that are impurities from some of the polymerization elements that were left over after the reaction was completed, as well as an impurity that could damage the system. While plastics' low density is generally advantageous for their application, gases and liquids can easily pass through them because of the relatively loose packing of their molecular framework. This has potential uses, especially in packing. Because certain plastics have extremely low permeabilities and some have high permeabilities, it is not possible to generalize about performance and degree of permeabilities as quantified in terms of the permeation constant (k). In particular, for materials that come into contact with food or are used in food packaging, there are proven qualitative and quantitative analytical techniques for characterizing migrants from plastics and rubbers. It was acknowledged a few years ago that migration modeling may be used by the plastics industry to assess plastic materials for compliance.

8.7. PROPER EVALUATION

It is also feasible to conduct a thorough environmental evaluation of packaging materials, taking into account both their life-cycle assessment and environmental impact. All polymers are poor electrical conductors and readily concentrate static charge on their surfaces, leading to a variety of issues such as consumer electrical shocks, and in certain extreme situations, spark-induced fire and explosions in industrial packaging. This is in addition to issues that arise when handling, storing, transporting, and processing delicate plastic electronic parts and gadgets, as well as the general issue of dust contamination, which can have an impact on the final product's functionality and appearance. These should all be minimized. The employment of intrinsically conducting polymers can make surfaces more conductive electrically, helping to dissipate high electrical charge densities and lowering the likelihood of any spark or discharge. Proper antistatic ESD agents can also be used to control or avoid catastrophic risks. Surface resistivities can be raised up to 50 times using ESD. An internal application is also preferred for high-quality applications, such as printing on surfaces. In this type of application, the hydrophilic end of the ESD agent projects out of the polymer, while its lipophilic end is anchored in the polymer. During processing, the internal non-permanent ESD migrates to the polymer's surface, building up a uniform layer. Internal ESDs are more effective and desirable, but because they are employed at much higher doses and consider migration probabilities, they may also be more harmful. In any event, ESD needs to meet all requirements and have the necessary food-contact certifications in order to be used in applications involving food.

By blending the polymer with an erythritol polymer fatty acid ester, it is possible to prepare Low-Density Polyethylene (LDPE) compositions with low volatile contents (Kato and Conte-Junior, 2021). Alternatively, specific agents can be added during processing to combine antistatic and fire-retardant properties. Given the widespread usage of antistatic agents in the system, it is typically required to secure specific and crucial permissions before using any of them, such as those from the US Food and Drug Administration (FDA). Since polymeric antistatic compounds have a high molecular weight and less migration in the bulk of the polymer, they can operate in polymeric systems for extended periods with increased safety and fewer health risks. They are still in the development stage, though. These polymeric antistatic agents, known as polyether polyamide copolymers, can safely give a variety of polymers permanent antistatic qualities. However, they are typically required in considerable concentrations, which may have an adverse effect on the mechanical properties.

8.8. COLORANTS-CONCERNS

Intentional additions called colorants are used to give plastic items color. Colorants are further classified as dispersed organic, inorganic, and special effect pigments, as well as dyes. Organic colorants known as dyes are readily soluble in polymers, either in the polymer itself or in one of the polymer components, and they have a tendency to migrate out of the system. Certain inorganic pigments that are used are derived from heavy metals such as barium, cadmium, iron, lead, mercury/chromium oxides, titanium, zinc, complex inorganic pigments that are mixtures of two or more metal oxides, and sulfurides, or they can be metals themselves such as aluminum, copper, and gold that are mixed with plastic and distributed as powders. The use of inorganic colors has drawn criticism since heavy metal ions can migrate out of the plastic and contaminate groundwater, even if the plastic is disposed of properly, posing a health risk. For these systems, incinerating waste made of colored plastic is not a solution since the ash contains metal residues. Hence, the use of any or all organic pigments in plastics intended for packaging is prohibited in a number of states. Wherever they are employed, their concentrations need to be far lower than 100 parts per million. They are capable of offering an optical shield against UV, IR, and visible light. Inorganic metallic pigments, such as lead, are employed to defend against nuclear radiation, especially at short wavelengths. They can enable the system to function as an antenna and microwave absorber for a variety of particular applications. Certain inorganic pigments are known to cause cancer. For example, pigments containing nickel are classified as Group 1 carcinogens; carbon black is classified as a Group 2B carcinogen; and pigments containing cobalt are classified as probably carcinogenic to humans (Group 1A), meaning that there is enough evidence to consider the substance as causing cancer in humans. Group 2B indicates a known carcinogen for animals and possibly a carcinogen in humans as well (Morita et al., 1997). Since there is no evidence that titanium dioxide, red iron oxide, or blue cobalt oxide inorganic pigments are harmful to human health, they are likewise included in the safe category. Zinc oxide is utilized in rubber, primarily as an accelerator, but it can also be employed directly as a crosslinker in polymers that include carboxylic or halogen.

CONCLUSION

These chapters summarized salient aspects of various plastics in terms of their degradation and effects on humans. Further research and careful use of these molecules can help safeguard their negative impacts on the environment.

BIBLIOGRAPHY

1. Aftalion, F. (2012). Recent history of the chemical industry 1973 to the millennium: The new facts of world chemicals since 1973. In J. A. Kent (Ed.), *Handbook of industrial chemistry and biotechnology* (pp. 1–45). Springer US. https://doi.org/10.1007/978-1-4614-4259-2_1.
2. Akdogan, Z., & Guven, B. (2019). Microplastics in the environment: A critical review of current understanding and identification of future research needs. *Environmental Pollution*, *254*, 113011. https://doi.org/10.1016/j.envpol.2019.113011.
3. Akelah, A. (2013). Polymers in food packaging and protection. In A. Akelah (Ed.), *Functionalized polymeric materials in agriculture and the food industry* (pp. 293–347). Springer US. https://doi.org/10.1007/978-1-4614-7061-8_6.
4. Ambrogi, V., Carfagna, C., Cerruti, P., & Marturano, V. (2017). Additives in polymers. In C. F. Jasso-Gastinel & J. M. Kenny (Eds.), *Modification of polymer properties* (pp. 87–108). William Andrew Publishing. https://doi.org/10.1016/B978-0-323-44353-1.00004-X.
5. Anbumani, S., & Kakkar, P. (2018). Ecotoxicological effects of microplastics on biota: A review. *Environmental Science and Pollution Research*, *25*(15), 14373–14396. https://doi.org/10.1007/s11356-018-1999-x.
6. Andrady, A. L. (2003). Common plastics materials. In *Plastics and the environment* (pp. 77–121). John Wiley & Sons, Ltd. https://doi.org/10.1002/0471721557.ch2.

7. Andrady, A. L., & Rajapakse, N. (2019). Additives and chemicals in plastics. In H. Takada & H. K. Karapanagioti (Eds.), *Hazardous chemicals associated with plastics in the marine environment* (pp. 1–17). Springer International Publishing. https://doi.org/10.1007/698_2016_124.

8. Andrady, A. L., Heikkilä, A. M., Pandey, K. K., Bruckman, L. S., White, C. C., Zhu, M., & Zhu, L. (2023). Effects of UV radiation on natural and synthetic materials. *Photochemical & Photobiological Sciences*, *22*(5), 1177–1202. https://doi.org/10.1007/s43630-023-00377-6.

9. Arévalo, J., Ruiz, L. M., Parada-Albarracín, J. A., González-Pérez, D. M., Pérez, J., Moreno, B., & Gómez, M. A. (2012). Wastewater reuse after treatment by MBR: Microfiltration or ultrafiltration? *Desalination*, *299*, 22–27. https://doi.org/10.1016/j.desal.2012.05.008.

10. Autrup, H., Barile, F. A., Berry, S. C., Blaauboer, B. J., Boobis, A., Bolt, H., Borgert, C. J., Dekant, W., Dietrich, D., Domingo, J. L., Gori, G. B., Greim, H., Hengstler, J., Kacew, S., Marquardt, H., Pelkonen, O., Savolainen, K., Heslop-Harrison, P., & Vermeulen, N. P. (2020). Human exposure to synthetic endocrine disrupting chemicals (S-EDCs) is generally negligible as compared to natural compounds with higher or comparable endocrine activity: How to evaluate the risk of the S-EDCs? *Toxicology in Vitro*, *67*, 104861. https://doi.org/10.1016/j.tiv.2020.104861.

11. Bajec, D., Grom, M., Lašič Jurković, D., Kostyniuk, A., Huš, M., Grilc, M., Likozar, B., & Pohar, A. (2020). A review of methane activation reactions by halogenation: Catalysis, mechanism, kinetics, modeling, and reactors. *Processes*, *8*(4), Article 4. https://doi.org/10.3390/pr8040443.

12. Batool, I., & Qadir, A. (2022). Atmospheric microplastic distribution, fate, and behavior in context to pollution. In M. Z. Hashmi (Ed.), *Microplastic pollution: Environmental occurrence and treatment technologies* (pp. 111–132). Springer International Publishing. https://doi.org/10.1007/978-3-030-89220-3_6.

13. Boyer, R. F. (1981). Anecdotal history of styrene and polystyrene. *Journal of Macromolecular Science: Part A - Chemistry*, *15*(7), 1411–1434. https://doi.org/10.1080/00222338108056791.

14. Bramwell, L., Harrad, S., Abou-Elwafa Abdallah, M., Rauert, C., Rose, M., Fernandes, A., & Pless-Mulloli, T. (2017). Predictors of human PBDE body burdens for a UK cohort. *Chemosphere*, *189*, 186–197. https://doi.org/10.1016/j.chemosphere.2017.08.062.

15. Carney Almroth, B. M., Åström, L., Roslund, S., Petersson, H., Johansson, M., & Persson, N.-K. (2018). Quantifying shedding of synthetic fibers from textiles; A source of microplastics released into the environment.

Environmental Science and Pollution Research, *25*(2), 1191–1199. https://doi.org/10.1007/s11356-017-0528-7.

16. Curtis, F. (2009). Peak globalization: Climate change, oil depletion and global trade. *Ecological Economics*, *69*(2), 427–434. https://doi.org/10.1016/j.ecolecon.2009.08.020.
17. da Costa, J. P., Duarte, A. C., & Costa, M. F. (2022). Collection and separation of microplastics. In T. Rocha-Santos, M. F. Costa, & C. Mouneyrac (Eds.), *Handbook of microplastics in the environment* (pp. 33–56). Springer International Publishing. https://doi.org/10.1007/978-3-030-39041-9_35.
18. De Tandt, E., Demuytere, C., Van Asbroeck, E., Moerman, H., Mys, N., Vyncke, G., Delva, L., Vermeulen, A., Ragaert, P., De Meester, S., & Ragaert, K. (2021). A recycler's perspective on the implications of REACH and food contact material (FCM) regulations for the mechanical recycling of FCM plastics. *Waste Management*, *119*, 315–329. https://doi.org/10.1016/j.wasman.2020.10.012.
19. den Braver-Sewradj, S. P., van Spronsen, R., & Hessel, E. V. S. (2020). Substitution of bisphenol A: A review of the carcinogenicity, reproductive toxicity, and endocrine disruption potential of alternative substances. *Critical Reviews in Toxicology*, *50*(2), 128–147. https://doi.org/10.1080/10408444.2019.1701986.
20. Desidery, L., & Lanotte, M. (2022). Polymers and plastics: Types, properties, and manufacturing. In F. Giustozzi & S. Nizamuddin (Eds.), *Plastic waste for sustainable asphalt roads* (pp. 3–28). Woodhead Publishing. https://doi.org/10.1016/B978-0-323-85789-5.00001-0.
21. Di, J., Reck, B. K., Miatto, A., & Graedel, T. E. (2021). United States plastics: Large flows, short lifetimes, and negligible recycling. *Resources, Conservation and Recycling*, *167*, 105440. https://doi.org/10.1016/j.resconrec.2021.105440.
22. Dimassi, S. N., Hahladakis, J. N., Chamkha, M., Ahmad, M. I., Al-Ghouti, M. A., & Sayadi, S. (2024). Investigation on the effect of several parameters involved in the biodegradation of polyethylene (PE) and low-density polyethylene (LDPE) under various seawater environments. *Science of The Total Environment*, *912*, 168870. https://doi.org/10.1016/j.scitotenv.2023.168870.
23. Dogu, O., Pelucchi, M., Van de Vijver, R., Van Steenberge, P. H. M., D'hooge, D. R., Cuoci, A., Mehl, M., Frassoldati, A., Faravelli, T., & Van Geem, K. M. (2021). The chemistry of chemical recycling of solid plastic waste via pyrolysis and gasification: State-of-the-art, challenges,

and future directions. *Progress in Energy and Combustion Science*, *84*, 100901. https://doi.org/10.1016/j.pecs.2020.100901.

24. Dong, M., Luo, Z., Jiang, Q., Xing, X., Zhang, Q., & Sun, Y. (2020). The rapid increases in microplastics in urban lake sediments. *Scientific Reports*, *10*(1), Article 1. https://doi.org/10.1038/s41598-020-57933-8.
25. Dumas, C., & Hsu, C. C. (1989). Propylene polymerization in a semibatch slurry reactor over supported TiCl4/MgCl2/ethyl benzoate/triethyl aluminum catalyst. I. Catalytic behavior. *Journal of Applied Polymer Science*, *37*(6), 1605–1623. https://doi.org/10.1002/app.1989.070370615.
26. Earls, A. O., Axford, I. P., & Braybrook, J. H. (2003). Gas chromatography–mass spectrometry determination of the migration of phthalate plasticisers from polyvinyl chloride toys and childcare articles. *Journal of Chromatography A*, *983*(1), 237–246. https://doi.org/10.1016/S0021-9673(02)01736-3.
27. Erickson, M. D., & Kaley, R. G. (2011). Applications of polychlorinated biphenyls. *Environmental Science and Pollution Research*, *18*(2), 135–151. https://doi.org/10.1007/s11356-010-0392-1.
28. Erythropel, H. C., Maric, M., Nicell, J. A., Leask, R. L., & Yargeau, V. (2014). Leaching of the plasticizer di(2-ethylhexyl)phthalate (DEHP) from plastic containers and the question of human exposure. *Applied Microbiology and Biotechnology*, *98*(24), 9967–9981. https://doi.org/10.1007/s00253-014-6183-8.
29. Fanini, L., Defeo, O., & Elliott, M. (2020). Advances in sandy beach research – Local and global perspectives. *Estuarine, Coastal and Shelf Science*, *234*, 106646. https://doi.org/10.1016/j.ecss.2020.106646.
30. Fantke, P., Weber, R., & Scheringer, M. (2015). From incremental to fundamental substitution in chemical alternatives assessment. *Sustainable Chemistry and Pharmacy*, *1*, 1–8. https://doi.org/10.1016/j.scp.2015.08.001.
31. Filho, W. L., Salvia, A. L., Bonoli, A., Saari, U. A., Voronova, V., Klõga, M., Kumbhar, S. S., Olszewski, K., De Quevedo, D. M., & Barbir, J. (2021). An assessment of attitudes towards plastics and bioplastics in Europe. *Science of The Total Environment*, *755*, 142732. https://doi.org/10.1016/j.scitotenv.2020.142732.
32. Fotopoulou, K. N., & Karapanagioti, H. K. (2019). Degradation of various plastics in the environment. In H. Takada & H. K. Karapanagioti (Eds.), *Hazardous chemicals associated with plastics in the marine environment* (pp. 71–92). Springer International Publishing. https://doi.org/10.1007/698_2017_11.

33. Fournier, E., Leveque, M., Ruiz, P., Ratel, J., Durif, C., Chalancon, S., Amiard, F., Edely, M., Bezirard, V., Gaultier, E., Lamas, B., Houdeau, E., Lagarde, F., Engel, E., Etienne-Mesmin, L., Blanquet-Diot, S., & Mercier-Bonin, M. (2023). Microplastics: What happens in the human digestive tract? First evidences in adults using in vitro gut models. *Journal of Hazardous Materials*, *442*, 130010. https://doi.org/10.1016/j.jhazmat.2022.130010.

34. Gao, S. Z., Wang, J. S., & Gao, X. W. (2013). Modeling and advanced control method of PVC polymerization process. *Journal of Process Control*, *23*(5), 664–681. https://doi.org/10.1016/j.jprocont.2013.02.008.

35. Geens, T., Aerts, D., Berthot, C., Bourguignon, J.-P., Goeyens, L., Lecomte, P., Maghuin-Rogister, G., Pironnet, A.-M., Pussemier, L., Scippo, M.-L., Van Loco, J., & Covaci, A. (2012). A review of dietary and non-dietary exposure to bisphenol-A. *Food and Chemical Toxicology*, *50*(10), 3725–3740. https://doi.org/10.1016/j.fct.2012.07.059.

36. Geens, T., Goeyens, L., & Covaci, A. (2011). Are potential sources for human exposure to bisphenol-A overlooked? *International Journal of Hygiene and Environmental Health*, *214*(5), 339–347. https://doi.org/10.1016/j.ijheh.2011.04.005.

37. Geyer, R. (2020). Production, use, and fate of synthetic polymers. In T. M. Letcher (Ed.), *Plastic waste and recycling* (pp. 13–32). Academic Press. https://doi.org/10.1016/B978-0-12-817880-5.00002-5.

38. Geyer, R., Jambeck, J. R., & Law, K. L. (2017). Production, use, and fate of all plastics ever made. *Science Advances*, *3*(7), e1700782. https://doi.org/10.1126/sciadv.1700782.

39. Gigault, J., El Hadri, H., Nguyen, B., Grassl, B., Rowenczyk, L., Tufenkji, N., Feng, S., & Wiesner, M. (2021). Nanoplastics are neither microplastics nor engineered nanoparticles. *Nature Nanotechnology*, *16*(5), Article 5. https://doi.org/10.1038/s41565-021-00886-4.

40. Godwin, A. D. (2017). Plasticizers. In M. Kutz (Ed.), *Applied plastics engineering handbook* (2nd ed., pp. 533–553). William Andrew Publishing. https://doi.org/10.1016/B978-0-323-39040-8.00025-0.

41. Gold, B. N. A., & Magaw, L. S. (1990). Ranking possible carcinogenic hazards. In *Readings in risk* (pp. 307–322). RFF Press.

42. Gong, M., Yang, G., Zhuang, L., & Zeng, E. Y. (2019). Microbial biofilm formation and community structure on low-density polyethylene microparticles in lake water microcosms. *Environmental Pollution*, *252*, 94–102. https://doi.org/10.1016/j.envpol.2019.05.090.

43. Graceli, J. B., Sena, G. C., Lopes, P. F. I., Zamprogno, G. C., da Costa, M.

B., Godoi, A. F. L., dos Santos, D. M., de Marchi, M. R. R., & dos Santos Fernandez, M. A. (2013). Organotins: A review of their reproductive toxicity, biochemistry, and environmental fate. *Reproductive Toxicology*, *36*, 40–52. https://doi.org/10.1016/j.reprotox.2012.11.008.

44. Hagens, N. J. (2020). Economics for the future – Beyond the superorganism. *Ecological Economics*, *169*, 106520. https://doi.org/10.1016/j.ecolecon.2019.106520.

45. Hamad, A., Khashan, K. S., & Hadi, A. (2020). Silver nanoparticles and silver ions as potential antibacterial agents. *Journal of Inorganic and Organometallic Polymers and Materials*, *30*(12), 4811–4828. https://doi.org/10.1007/s10904-020-01744-x.

46. He, Z.-W., Yang, W.-J., Ren, Y.-X., Jin, H.-Y., Tang, C.-C., Liu, W.-Z., Yang, C.-X., Zhou, A.-J., & Wang, A.-J. (2021). Occurrence, effect, and fate of residual microplastics in anaerobic digestion of waste activated sludge: A state-of-the-art review. *Bioresource Technology*, *331*, 125035. https://doi.org/10.1016/j.biortech.2021.125035.

47. Hermabessiere, L., Dehaut, A., Paul-Pont, I., Lacroix, C., Jezequel, R., Soudant, P., & Duflos, G. (2017). Occurrence and effects of plastic additives on marine environments and organisms: A review. *Chemosphere*, *182*, 781–793. https://doi.org/10.1016/j.chemosphere.2017.05.096.

48. Herrera, A., Garrido-Amador, P., Martínez, I., Samper, M. D., López-Martínez, J., Gómez, M., & Packard, T. T. (2018). Novel methodology to isolate microplastics from vegetal-rich samples. *Marine Pollution Bulletin*, *129*(1), 61–69. https://doi.org/10.1016/j.marpolbul.2018.02.015.

49. Hohl, H., & Varma, A. (2010). Soil: The living matrix. In *Soil heavy metals* (pp. 1–18). Springer. https://doi.org/10.1007/978-3-642-02436-8_1.

50. Hou, D., Hong, M., Wang, Y., Dong, P., Cheng, H., Yan, H., Yao, Z., Li, D., Wang, K., & Zhang, D. (2021). Assessing the risks of potential bacterial pathogens attaching to different microplastics during the summer–autumn period in a mariculture cage. *Microorganisms*, *9*(9), Article 9. https://doi.org/10.3390/microorganisms9091909.

51. Hüffer, T., & Hofmann, T. (2016). Sorption of non-polar organic compounds by micro-sized plastic particles in aqueous solution. *Environmental Pollution*, *214*, 194–201. https://doi.org/10.1016/j.envpol.2016.04.018.

52. Hutley, T. J., & Ouederni, M. (2016). Polyolefins—The history and economic impact. In M. Al-Ali AlMa'adeed & I. Krupa (Eds.), *Polyolefin compounds and materials: Fundamentals and industrial applications* (pp. 13–50). Springer International Publishing. https://doi.org/10.1007/978-3-319-25982-6_2.

53. Idumah, C. I., Hassan, A., & Ihuoma, D. E. (2019). Recently emerging trends in polymer nanocomposites packaging materials. *Polymer-Plastics Technology and Materials*, *58*(10), 1054–1109. https://doi.org/10.1080/03602559.2018.1542718.

54. Ioakeimidis, C., Galgani, F., & Papatheodorou, G. (2019). Occurrence of marine litter in the marine environment: A world panorama of floating and seafloor plastics. In H. Takada & H. K. Karapanagioti (Eds.), *Hazardous chemicals associated with plastics in the marine environment* (pp. 93–120). Springer International Publishing. https://doi.org/10.1007/698_2017_22.

55. Issac, M. N., & Kandasubramanian, B. (2021). Effect of microplastics in water and aquatic systems. *Environmental Science and Pollution Research*, *28*(16), 19544–19562. https://doi.org/10.1007/s11356-021-13184-2.

56. Istrate, I.-R., Juan, R., Martin-Gamboa, M., Domínguez, C., García-Muñoz, R. A., & Dufour, J. (2021). Environmental life cycle assessment of the incorporation of recycled high-density polyethylene to polyethylene pipe grade resins. *Journal of Cleaner Production*, *319*, 128580. https://doi.org/10.1016/j.jclepro.2021.128580.

57. Jagadeesh, P., Mavinkere Rangappa, S., Siengchin, S., Puttegowda, M., Thiagamani, S. M. K., G., R., Hemath Kumar, M., Oladijo, O. P., Fiore, V., & Moure Cuadrado, M. M. (2022). Sustainable recycling technologies for thermoplastic polymers and their composites: A review of the state of the art. *Polymer Composites*, *43*(9), 5831–5862. https://doi.org/10.1002/pc.27000.

58. Jahan, K., Hoque, S., & Ahmed, T. (2012). Activated sludge and other suspended culture processes. *Water Environment Research*, *84*(10), 1029–1080. https://doi.org/10.2175/106143012X13407275694798.

59. Jiang, X., & He, A. (2014). Stereospecific polymerization of olefins with supported Ziegler − Natta catalysts. *Polymer International*, *63*(2), 179–183. https://doi.org/10.1002/pi.4633.

60. Johnson, K. S., Coale, K. H., & Jannasch, H. W. (2008, September 22). Analytical chemistry and oceanography (world) [Research-article]. ACS Publications; American Chemical Society. https://doi.org/10.1021/ac00046a001.

61. Ju, H., Yang, X., Osman, R., & Geissen, V. (2023). Effects of microplastics and chlorpyrifos on earthworms (*Lumbricus terrestris*) and their biogenic transport in sandy soil. *Environmental Pollution*, *316*, 120483. https://doi.org/10.1016/j.envpol.2022.120483.

62. Kalair, A., Abas, N., Saleem, M. S., Kalair, A. R., & Khan, N. (2021). Role of energy storage systems in energy transition from fossil fuels to

renewables. *Energy Storage*, *3*(1), e135. https://doi.org/10.1002/est2.135.

63. Kamrin, M. A. (2009). Phthalate risks, phthalate regulation, and public health: A review. *Journal of Toxicology and Environmental Health, Part B*, *12*(2), 157–174. https://doi.org/10.1080/10937400902729226.

64. Karol, F. J. (1986). Catalysis and the polyethylene revolution. In R. B. Seymour & T. Cheng (Eds.), *History of polyolefins: The world's most widely used polymers* (pp. 193–211). Springer Netherlands. https://doi.org/10.1007/978-94-009-5472-4_12.

65. Kato, L. S., & Conte-Junior, C. A. (2021). Safety of plastic food packaging: The challenges about non-intentionally added substances (NIAS) discovery, identification and risk assessment. *Polymers*, *13*(13), Article 13. https://doi.org/10.3390/polym13132077.

66. Kedzierski, M., Le Tilly, V., Bourseau, P., Bellegou, H., César, G., Sire, O., & Bruzaud, S. (2016). Microplastics elutriation from sandy sediments: A granulometric approach. *Marine Pollution Bulletin*, *107*(1), 315–323. https://doi.org/10.1016/j.marpolbul.2016.03.041.

67. Keresztes, S., Tatár, E., Mihucz, V. G., Virág, I., Majdik, C., & Záray, G. (2009). Leaching of antimony from polyethylene terephthalate (PET) bottles into mineral water. *Science of The Total Environment*, *407*(16), 4731–4735. https://doi.org/10.1016/j.scitotenv.2009.04.025.

68. Khadir, A., Negarestani, M., Kheradmand, A., & Sillanpää, M. (2022). Microplastic pollution in water and their removal in various wastewater treatment plants. In M. Sillanpää, A. Khadir, & S. S. Muthu (Eds.), *Microplastics pollution in aquatic media: Occurrence, detection, and removal* (pp. 247–271). Springer. https://doi.org/10.1007/978-981-16-8440-1_11.

69. Kiparissides, C., Verros, G., & Macgregor, J. F. (1993). Mathematical modeling, optimization, and quality control of high-pressure ethylene polymerization reactors. *Journal of Macromolecular Science, Part C*, *33*(4), 437–527. https://doi.org/10.1080/15321799308021566.

70. Knuuttila, H., Lehtinen, A., & Nummila-Pakarinen, A. (2004). Advanced polyethylene technologies—Controlled material properties. In A.-C. Albertsson (Ed.), *Long term properties of polyolefins* (pp. 13–28). Springer. https://doi.org/10.1007/b13519.

71. Kopatz, V., Wen, K., Kovács, T., Keimowitz, A. S., Pichler, V., Widder, J., Vethaak, A. D., Hollóczki, O., & Kenner, L. (2023). Micro- and nanoplastics breach the blood–brain barrier (BBB): Biomolecular corona's role revealed. *Nanomaterials*, *13*(8), Article 8. https://doi.org/10.3390/nano13081404.

72. Krewski, D., Acosta Jr., D., Andersen, M., Anderson, H., Bailar III, J. C., Boekelheide, K., Brent, R., Charnley, G., Cheung, V. G., Green Jr., S., Kelsey, K. T., Kerkvliet, N. I., Li, A. A., McCray, L., Meyer, O., Patterson, R. D., Pennie, W., Scala, R. A., Solomon, G. M., … Staff of Committee on Toxicity Testing and Assessment of Environmental Agents. (2010). Toxicity testing in the 21st century: A vision and a strategy. *Journal of Toxicology and Environmental Health, Part B*, *13*(2–4), 51–138. https://doi.org/10.1080/10937404.2010.483176.

73. Kuczenski, B., & Geyer, R. (2010). Material flow analysis of polyethylene terephthalate in the US, 1996–2007. *Resources, Conservation and Recycling*, *54*(12), 1161–1169. https://doi.org/10.1016/j.resconrec.2010.03.013.

74. Lebreton, L., & Andrady, A. (2019). Future scenarios of global plastic waste generation and disposal. *Palgrave Communications*, *5*(1), Article 1. https://doi.org/10.1057/s41599-018-0212-7.

75. Li, L., Li, Z., Liu, D., & Song, K. (2020). Evaluation of partial nitrification efficiency as a response to cadmium concentration and microplastic polyvinylchloride abundance during landfill leachate treatment. *Chemosphere*, *247*, 125903. https://doi.org/10.1016/j.chemosphere.2020.125903.

76. Li, Y., Liu, C., Yang, H., He, W., Li, B., Zhu, X., Liu, S., Jia, S., Li, R., & Tang, K. H. D. (2024). Leaching of chemicals from microplastics: A review of chemical types, leaching mechanisms and influencing factors. *Science of The Total Environment*, *906*, 167666. https://doi.org/10.1016/j.scitotenv.2023.167666.

77. Lin, X., Su, C., Deng, X., Wu, S., Tang, L., Li, X., Liu, J., & Huang, X. (2020). Influence of polyether sulfone microplastics and bisphenol A on anaerobic granular sludge: Performance evaluation and microbial community characterization. *Ecotoxicology and Environmental Safety*, *205*, 111318. https://doi.org/10.1016/j.ecoenv.2020.111318.

78. Liu, X., Miao, M., Zhou, Z., Gao, E., Chen, J., Wang, J., Sun, F., Yuan, W., & Li, D.-K. (2015). Exposure to bisphenol-A and reproductive hormones among male adults. *Environmental Toxicology and Pharmacology*, *39*(2), 934–941. https://doi.org/10.1016/j.etap.2015.03.007.

79. Luo, H., Zeng, Y., Cheng, Y., He, D., & Pan, X. (2020). Recent advances in municipal landfill leachate: A review focusing on its characteristics, treatment, and toxicity assessment. *Science of The Total Environment*, *703*, 135468. https://doi.org/10.1016/j.scitotenv.2019.135468.

80. MacDonald, W. A. (2015). Latest advances in substrates for flexible electronics. In *Large area and flexible electronics* (pp. 291–314). John

Wiley & Sons, Ltd. https://doi.org/10.1002/9783527679973.ch10.

81. Mai, L., Sun, X.-F., Xia, L.-L., Bao, L.-J., Liu, L.-Y., & Zeng, E. Y. (2020). Global riverine plastic outflows. *Environmental Science & Technology, 54*(16), 10049–10056. https://doi.org/10.1021/acs.est.0c02273.

82. Marcinčin, A. (2002). Modification of fiber-forming polymers by additives. *Progress in Polymer Science, 27*(5), 853–913. https://doi.org/10.1016/S0079-6700(02)00002-3.

83. McCarthy, M. M., & Ball, G. F. (2008). The neuroendocrine control of sex specific behavior in vertebrates: Lessons from mammals and birds. In *Current topics in developmental biology* (Vol. 83, pp. 213–248). Academic Press. https://doi.org/10.1016/S0070-2153(08)00407-9.

84. Meister, B. J., & Cummings, C. J. (2003). Commercial processes for the manufacture of polystyrene. In *Modern styrenic polymers: Polystyrenes and styrenic copolymers* (pp. 43–72). John Wiley & Sons, Ltd. https://doi.org/10.1002/0470867213.ch3.

85. Mhaske, S., Chugh, K., & Mahajan, U. (2023). Types of polymeric foams. In *Polymeric foams: Fundamentals and types of foams* (Vol. 1, pp. 25–41). American Chemical Society. https://doi.org/10.1021/bk-2023-1439.ch002.

86. Mokwena, K. K., & Tang, J. (2012). Ethylene vinyl alcohol: A review of barrier properties for packaging shelf stable foods. *Critical Reviews in Food Science and Nutrition, 52*(7), 640–650. https://doi.org/10.1080/10408398.2010.504903.

87. Moore, C. J. (2008). Synthetic polymers in the marine environment: A rapidly increasing, long-term threat. *Environmental Research, 108*(2), 131–139. https://doi.org/10.1016/j.envres.2008.07.025.

88. Morita, T., Asano, N., Awogi, T., Sasaki, Y. F., Sato, S.-I., Shimada, H., Sutou, S., Suzuki, T., Wakata, A., Sofuni, T., & Hayashi, M. (1997). Evaluation of the rodent micronucleus assay in the screening of IARC carcinogens (Groups 1, 2A, and 2B): The summary report of the 6th collaborative study by CSGMT/JEMS·MMS. *Mutation Research/Genetic Toxicology and Environmental Mutagenesis, 389*(1), 3–122. https://doi.org/10.1016/S1383-5718(96)00070-8.

89. Mosquera-Losada, R., Amador-García, A., Muñóz-Ferreiro, N., Santiago-Freijanes, J. J., Ferreiro-Domínguez, N., Romero-Franco, R., & Rigueiro-Rodríguez, A. (2017). Sustainable use of sewage sludge in acid soils within a circular economy perspective. *CATENA, 149*, 341–348. https://doi.org/10.1016/j.catena.2016.10.007.

90. Muncke, J. (2011). Endocrine disrupting chemicals and other substances of

concern in food contact materials: An updated review of exposure, effect, and risk assessment. *The Journal of Steroid Biochemistry and Molecular Biology*, *127*(1), 118–127. https://doi.org/10.1016/j.jsbmb.2010.10.004.

91. Nakajima, H., Dijkstra, P., & Loos, K. (2017). The recent developments in biobased polymers toward general and engineering applications: Polymers that are upgraded from biodegradable polymers, analogous to petroleum-derived polymers, and newly developed. *Polymers*, *9*(10), Article 10. https://doi.org/10.3390/polym9100523.
92. Nguyen, B., Claveau-Mallet, D., Hernandez, L. M., Xu, E. G., Farner, J. M., & Tufenkji, N. (2019). Separation and analysis of microplastics and nanoplastics in complex environmental samples. *Accounts of Chemical Research*, *52*(4), 858–866. https://doi.org/10.1021/acs.accounts.8b00602.
93. Nielsen, T. D., Holmberg, K., & Stripple, J. (2019). Need a bag? A review of public policies on plastic carrier bags – Where, how, and to what effect? *Waste Management*, *87*, 428–440. https://doi.org/10.1016/j.wasman.2019.02.025.
94. Niessner, N., & Gausepohl, H. (2003). Polystyrenes and styrene copolymers – An overview. In *Modern styrenic polymers: Polystyrenes and styrenic copolymers* (pp. 25–42). John Wiley & Sons, Ltd. https://doi.org/10.1002/0470867213.ch2.
95. Norton, M. G. (2023). A Green New Deal. In *A modern history of materials: From stability to sustainability* (pp. 187–206). Springer International Publishing. https://doi.org/10.1007/978-3-031-23990-8_10.
96. Odzak, N., Kistler, D., Behra, R., & Sigg, L. (2014). Dissolution of metal and metal oxide nanoparticles in aqueous media. *Environmental Pollution*, *191*, 132–138. https://doi.org/10.1016/j.envpol.2014.04.010.
97. Patel, R. M. (2017). Types and basics of polyethylene. In *Handbook of industrial polyethylene and technology* (pp. 105–138). John Wiley & Sons, Ltd. https://doi.org/10.1002/9781119159797.ch4.
98. Peeters, J. R., Vanegas, P., Tange, L., Van Houwelingen, J., & Duflou, J. R. (2014). Closed loop recycling of plastics containing flame retardants. *Resources, Conservation and Recycling*, *84*, 35–43. https://doi.org/10.1016/j.resconrec.2013.12.006.
99. Pillai, K. V., Hunt, P. R., & Duncan, T. V. (2014). Nanoparticles in polymer nanocomposite food contact materials: Uses, potential release, and emerging toxicological concerns. In S. M. Snedeker (Ed.), *Toxicants in food packaging and household plastics: Exposure and health risks to consumers* (pp. 95–123). Springer. https://doi.org/10.1007/978-1-4471-6500-2_4.

100. Prochazka, M. (2016). Basics of Raman scattering (RS) spectroscopy. In *Surface-enhanced Raman spectroscopy: Bioanalytical, biomolecular and medical applications* (pp. 7–19). Springer International Publishing. https://doi.org/10.1007/978-3-319-23992-7_2.

101. Raju, S., Carbery, M., Kuttykattil, A., Senthirajah, K., Lundmark, A., Rogers, Z., Scb, S., Evans, G., & Palanisami, T. (2020). Improved methodology to determine the fate and transport of microplastics in a secondary wastewater treatment plant. *Water Research, 173*, 115549. https://doi.org/10.1016/j.watres.2020.115549.

102. Razeghi, N., Hamidian, A. H., Wu, C., Zhang, Y., & Yang, M. (2021). Microplastic sampling techniques in freshwaters and sediments: A review. *Environmental Chemistry Letters*, *19*(6), 4225–4252. https://doi.org/10.1007/s10311-021-01227-6.

103. Reimann, S., Elkins, J. W., Fraser, P. J., Hall, B. D., Kurylo, M. J., Mahieu, E., Montzka, S. A., Prinn, R. G., Rigby, M., Simmonds, P. G., & Weiss, R. F. (2018). Observing the atmospheric evolution of ozone-depleting substances. *Comptes Rendus Geoscience*, *350*(7), 384–392. https://doi.org/10.1016/j.crte.2018.08.008.

104. Rios, L. M., Moore, C., & Jones, P. R. (2007). Persistent organic pollutants carried by synthetic polymers in the ocean environment. *Marine Pollution Bulletin*, *54*(8), 1230–1237. https://doi.org/10.1016/j.marpolbul.2007.03.022.

105. Rodgers, K. M., Rudel, R. A., & Just, A. C. (2014). Phthalates in food packaging, consumer products, and indoor environments. In S. M. Snedeker (Ed.), *Toxicants in food packaging and household plastics: Exposure and health risks to consumers* (pp. 31–59). Springer. https://doi.org/10.1007/978-1-4471-6500-2_2.

106. Ronca, S. (2017). Polyethylene. In M. Gilbert (Ed.), *Brydson's plastics materials* (8th ed., pp. 247–278). Butterworth-Heinemann. https://doi.org/10.1016/B978-0-323-35824-8.00010-4.

107. Ryan, P. G. (2019). Ingestion of plastics by marine organisms. In H. Takada & H. K. Karapanagioti (Eds.), *Hazardous chemicals associated with plastics in the marine environment* (pp. 235–266). Springer International Publishing. https://doi.org/10.1007/698_2016_21.

108. Samuels, R. J. (1969). Solid-state characterization of the structure and deformation behavior of water-soluble hydroxypropylcellulose. *Journal of Polymer Science Part A-2: Polymer Physics*, *7*(7), 1197–1258. https://doi.org/10.1002/pol.1969.160070705.

109. Scherbov, W., Sanderson, W. C., & Ospina, S. (2004). The end of world

population growth. In *The end of world population growth in the 21st century* (pp. 27–49). Routledge.

110. Schiaparelli, S., Schnabel, K. E., de Forges, B. R., & Chan, T.-Y. (2016). Sorting, recording, preservation, and storage of biological samples. In *Biological sampling in the deep sea* (pp. 338–367). John Wiley & Sons, Ltd. https://doi.org/10.1002/9781118332535.ch15.
111. Schymanski, D., Oßmann, B. E., Benismail, N., Boukerma, K., Dallmann, G., von der Esch, E., Fischer, D., Fischer, F., Gilliland, D., Glas, K., Hofmann, T., Käppler, A., Lacorte, S., Marco, J., Rakwe, M. E., Weisser, J., Witzig, C., Zumbülte, N., & Ivleva, N. P. (2021). Analysis of microplastics in drinking water and other clean water samples with micro-Raman and micro-infrared spectroscopy: Minimum requirements and best practice guidelines. *Analytical and Bioanalytical Chemistry*, *413*(24), 5969–5994. https://doi.org/10.1007/s00216-021-03498-y.
112. Seyfullah, L. J., Beimforde, C., Dal Corso, J., Perrichot, V., Rikkinen, J., & Schmidt, A. R. (2018). Production and preservation of resins – past and present. *Biological Reviews*, *93*(3), 1684–1714. https://doi.org/10.1111/brv.12414.
113. Seymour, R. B., Mark, H. F., Pauling, L., Fisher, C. H., Stahl, G. A., Sperling, L. H., Marvel, C. S., & Carraher, C. E. (1989). The development of thermosets by Lee Baekeland and other early 20th century chemists. In *Pioneers in polymer science* (pp. 81–92). Springer Netherlands. https://doi.org/10.1007/978-94-009-2407-9_7.
114. Shah, A. A., Hasan, F., Hameed, A., & Ahmed, S. (2008). Biological degradation of plastics: A comprehensive review. *Biotechnology Advances*, *26*(3), 246–265. https://doi.org/10.1016/j.biotechadv.2007.12.005.
115. Shaw, S. D., Harris, J. H., Berger, M. L., Subedi, B., & Kannan, K. (2014). Brominated flame retardants and their replacements in food packaging and household products: Uses, human exposure, and health effects. In S. M. Snedeker (Ed.), *Toxicants in food packaging and household plastics: Exposure and health risks to consumers* (pp. 61–93). Springer. https://doi.org/10.1007/978-1-4471-6500-2_3.
116. Shen, J., Kang, Q., Mao, Y., Yuan, M., Le, F., Yang, X., Xu, X., & Jin, F. (2020). Urinary bisphenol A concentration is correlated with poorer oocyte retrieval and embryo implantation outcomes in patients with tubal factor infertility undergoing in vitro fertilisation. *Ecotoxicology and Environmental Safety*, *187*, 109816. https://doi.org/10.1016/j.ecoenv.2019.109816.
117. Sikdar, S., Siddaiah, A., & Menezes, P. L. (2020). Conversion of waste

plastic to oils for tribological applications. *Lubricants*, *8*(8), Article 8. https://doi.org/10.3390/lubricants8080078.

118. Singh, R. L., & Singh, P. K. (2017). Global environmental problems. In R. L. Singh (Ed.), *Principles and applications of environmental biotechnology for a sustainable future* (pp. 13–41). Springer. https://doi.org/10.1007/978-981-10-1866-4_2.
119. Smith, J. K. (1988). World War II and the transformation of the American chemical industry. In E. Mendelsohn, M. R. Smith, & P. Weingart (Eds.), *Science, technology and the military* (pp. 307–322). Springer Netherlands. https://doi.org/10.1007/978-94-017-2958-1_2.
120. Stubbings, W. A., & Harrad, S. (2014). Extent and mechanisms of brominated flame retardant emissions from waste soft furnishings and fabrics: A critical review. *Environment International*, *71*, 164–175. https://doi.org/10.1016/j.envint.2014.06.007.
121. Sun, X., Anoopkumar, A. N., Madhavan, A., Binod, P., Pandey, A., Sindhu, R., & Awasthi, M. K. (2023). Degradation mechanism of microplastics and potential risks during sewage sludge co-composting: A comprehensive review. *Environmental Pollution*, *333*, 122113. https://doi.org/10.1016/j.envpol.2023.122113.
122. Sun, Y., Yuan, J., Zhou, T., Zhao, Y., Yu, F., & Ma, J. (2020). Laboratory simulation of microplastics weathering and its adsorption behaviors in an aqueous environment: A systematic review. *Environmental Pollution*, *265*, 114864. https://doi.org/10.1016/j.envpol.2020.114864.
123. Swinnen, J., Maertens, M., & Colen, L. (2015). The role of food standards in trade and development. In A. Hammoudi, C. Grazia, Y. Surry, & J.-B. Traversac (Eds.), *Food safety, market organization, trade and development* (pp. 133–149). Springer International Publishing. https://doi.org/10.1007/978-3-319-15227-1_7.
124. Syranidou, E., & Kalogerakis, N. (2022). Interactions of microplastics, antibiotics and antibiotic resistant genes within WWTPs. *Science of The Total Environment*, *804*, 150141. https://doi.org/10.1016/j.scitotenv.2021.150141.
125. Thiagarajan, V., Alex, S. A., Seenivasan, R., Chandrasekaran, N., & Mukherjee, A. (2021). Interactive effects of micro/nanoplastics and nanomaterials/pharmaceuticals: Their ecotoxicological consequences in the aquatic systems. *Aquatic Toxicology*, *232*, 105747. https://doi.org/10.1016/j.aquatox.2021.105747.
126. Tian, W., Song, P., Zhang, H., Duan, X., Wei, Y., Wang, H., & Wang, S. (2023). Microplastic materials in the environment: Problem and

strategical solutions. *Progress in Materials Science*, *132*, 101035. https://doi.org/10.1016/j.pmatsci.2022.101035.

127. Turner, A., & Filella, M. (2021). Hazardous metal additives in plastics and their environmental impacts. *Environment International*, *156*, 106622. https://doi.org/10.1016/j.envint.2021.106622.
128. Utracki, L. A. (1998). Polymer industry. In L. A. Utracki (Ed.), *Commercial polymer blends* (pp. 8–52). Springer US. https://doi.org/10.1007/978-1-4615-5789-0_2.
129. van Wijnen, J., Ragas, A. M. J., & Kroeze, C. (2019). Modelling global river export of microplastics to the marine environment: Sources and future trends. *Science of The Total Environment*, *673*, 392–401. https://doi.org/10.1016/j.scitotenv.2019.04.078.
130. Vilarinho, F., Sendón, R., van der Kellen, A., Vaz, M. F., & Silva, A. S. (2019). Bisphenol A in food as a result of its migration from food packaging. *Trends in Food Science & Technology*, *91*, 33–65. https://doi.org/10.1016/j.tifs.2019.06.012.
131. vom Saal, F. S., & Welshons, W. V. (2014). Evidence that bisphenol A (BPA) can be accurately measured without contamination in human serum and urine, and that BPA causes numerous hazards from multiple routes of exposure. *Molecular and Cellular Endocrinology*, *398*(1), 101–113. https://doi.org/10.1016/j.mce.2014.09.028.
132. Wang, F., Zhang, M., Sha, W., Wang, Y., Hao, H., Dou, Y., & Li, Y. (2020). Sorption behavior and mechanisms of organic contaminants to nano and microplastics. *Molecules*, *25*(8), Article 8. https://doi.org/10.3390/molecules25081827.
133. Wang, R., Wang, Q., Ma, C., Li, S., & Han, R. (2018). Phthalates in soft glass (a soft transparent PVC plastic sheet used extensively in household and public place in developing countries in recent years): Implication for oral exposure to young children. *Chemosphere*, *211*, 861–866. https://doi.org/10.1016/j.chemosphere.2018.08.002.
134. Wang, W., & Wang, J. (2018). Investigation of microplastics in aquatic environments: An overview of the methods used, from field sampling to laboratory analysis. *TrAC Trends in Analytical Chemistry*, *108*, 195–202. https://doi.org/10.1016/j.trac.2018.08.026.
135. Wang, X., Zhang, Y., Zhao, Y., Zhang, L., & Zhang, X. (2023). Inhibition of aged microplastics and leachates on methane production from anaerobic digestion of sludge and identification of key components. *Journal of Hazardous Materials*, *446*, 130717. https://doi.org/10.1016/j.jhazmat.2022.130717.

136. Wang, Y., Zhao, W., Wang, X., & Wu, D. (2016). Preparation, mechanical properties and microstructure of polyoxymethylene fiber through melt spinning and hot drawing by using injection-molding grade resins. *Fibers and Polymers*, *17*(9), 1464–1474. https://doi.org/10.1007/s12221-016-6586-5.

137. Wei, W., Chen, X., Peng, L., Liu, Y., Bao, T., & Ni, B.-J. (2021). The entering of polyethylene terephthalate microplastics into biological wastewater treatment system affects aerobic sludge digestion differently from their direct entering into sludge treatment system. *Water Research*, *190*, 116731. https://doi.org/10.1016/j.watres.2020.116731.

138. Williams, E. S., Panko, J., & Paustenbach, D. J. (2009). The European Union's REACH regulation: A review of its history and requirements. *Critical Reviews in Toxicology*, *39*(7), 553–575. https://doi.org/10.1080/10408440903036056.

139. Wolff, S., Weber, F., Kerpen, J., Winklhofer, M., Engelhart, M., & Barkmann, L. (2021). Elimination of microplastics by downstream sand filters in wastewater treatment. *Water*, *13*(1), Article 1. https://doi.org/10.3390/w13010033.

140. Wu, P., Huang, J., Zheng, Y., Yang, Y., Zhang, Y., He, F., Chen, H., Quan, G., Yan, J., Li, T., & Gao, B. (2019). Environmental occurrences, fate, and impacts of microplastics. *Ecotoxicology and Environmental Safety*, *184*, 109612. https://doi.org/10.1016/j.ecoenv.2019.109612.

141. Xu, Z., Bai, X., & Ye, Z. (2021). Removal and generation of microplastics in wastewater treatment plants: A review. *Journal of Cleaner Production*, *291*, 125982. https://doi.org/10.1016/j.jclepro.2021.125982.

142. Yates, J. G., & Lettieri, P. (2016). Fluidized-bed reactors: Processes and operating conditions (Vol. 26). Springer International Publishing. https://doi.org/10.1007/978-3-319-39593-7.

143. You, H. H., & Song, G. (2021). Review of endocrine disruptors on male and female reproductive systems. *Comparative Biochemistry and Physiology Part C: Toxicology & Pharmacology*, *244*, 109002. https://doi.org/10.1016/j.cbpc.2021.109002.

144. Yu, B., Zhao, T., Gustave, W., Li, B., Cai, Y., Ouyang, D., Guo, T., & Zhang, H. (2023). Do microplastics affect sulfamethoxazole sorption in soil? Experiments on polymers, ionic strength, and fulvic acid. *Science of The Total Environment*, *860*, 160221. https://doi.org/10.1016/j.scitotenv.2022.160221.

145. Yu, F., Yang, C., Zhu, Z., Bai, X., & Ma, J. (2019). Adsorption behavior of organic pollutants and metals on micro/nanoplastics in the aquatic

environment. *Science of The Total Environment*, *694*, 133643. https://doi.org/10.1016/j.scitotenv.2019.133643.

146. Zakharychev, V. V., Kuzenkov, A. V., & Martsynkevich, A. M. (2020). Good pyridine hunting: A biomimic compound, a modifier and a unique pharmacophore in agrochemicals. *Chemistry of Heterocyclic Compounds*, *56*(12), 1491–1516. https://doi.org/10.1007/s10593-020-02843-w.
147. Zhang, H. (2017). Transport of microplastics in coastal seas. *Estuarine, Coastal and Shelf Science*, *199*, 74–86. https://doi.org/10.1016/j.ecss.2017.09.032.
148. Zhang, Y.-T., Wei, W., Wang, C., & Ni, B.-J. (2022). Microbial and physicochemical responses of anaerobic hydrogen-producing granular sludge to polyethylene micro(nano)plastics. *Water Research*, *221*, 118745. https://doi.org/10.1016/j.watres.2022.118745.
149. Zhang, Y., Diehl, A., Lewandowski, A., Gopalakrishnan, K., & Baker, T. (2020). Removal efficiency of micro- and nanoplastics (180 nm–125 μm) during drinking water treatment. *Science of The Total Environment*, *720*, 137383. https://doi.org/10.1016/j.scitotenv.2020.137383.
150. Zhang, Z. L., Leith, C., Rhind, S. M., Kerr, C., Osprey, M., Kyle, C., Coull, M., Thomson, C., Green, G., Maderova, L., & McKenzie, C. (2014). Long term temporal and spatial changes in the distribution of polychlorinated biphenyls and polybrominated diphenyl ethers in Scottish soils. *Science of The Total Environment*, *468–469*, 158–164. https://doi.org/10.1016/j.scitotenv.2013.08.029.
151. Zhang, Z., & Chen, Y. (2020). Effects of microplastics on wastewater and sewage sludge treatment and their removal: A review. *Chemical Engineering Journal*, *382*, 122955. https://doi.org/10.1016/j.cej.2019.122955.
152. Zhao, W., & Schroeder, P. (2010). Sustainable consumption and production: Trends, challenges and options for the Asia-Pacific region. *Natural Resources Forum*, *34*(1), 4–15. https://doi.org/10.1111/j.1477-8947.2010.01275.x.

INDEX

A

Absorption process 171
Acetate Kinase (AK) 160
Acrylonitrile-Butadiene-Styrene (ABS) 132
Acrylonitrile–Butadiene–Styrene (ABS) 23
Activated Sludge Process (ASP) 146
Active sampling techniques 39
Additive mixture 26
Additive sales 82
Aerobic conditions 146
Agricultural soils 38, 75, 103
Airborne fallout 32
Alkali metal oxides 85
Alkylphenol Ethoxylates (APE) 175
Alpine Glaciers 37
Amoxicillin (AMX) 164
Anaerobic-anoxic-aerobic 146
Animal sampling 68
Antarctic 32
Anti-Stokes 49, 51
Aquatic environment 57, 70, 107, 112, 118, 143, 145, 162, 165, 166
Aquatic media 108, 116
Aquatic organisms 107
Artificial radiation 97
Aryl hydrocarbon receptor (AhR) 123
Atmospheric deposition 37, 60, 103
Atmospheric transport 38
Attention-Deficit/Hyperactivity Disorder (ADHD) 128
Attenuated Total Reflectance (ATR) 111
Autism Spectrum Disorders (ASDs) 128

B

Backwashing membrane filters 144
Bacteria 149, 166
Beach Sediment (BS) 73
Behavioral abnormalities 114, 115, 123
Biaxially Oriented Polypropylene film (BOPP) 19
Bioaccumulate 82, 124, 175, 176
Bio-accumulate phthalates 82
Bioaccumulation 44, 63, 80, 82, 83,

102
Bioavailable environmental pool 21
Biochemical Oxygen Demand (BOD) 147
Biodegrade 57, 162
Biodiversity 5, 70
Biological Active Filters (BAF) 147
Biological filter 146, 147
Biological matrices 129
Biological Treatment 155
Biomonitoring 97, 124, 128, 130
Biowaste 38
Bisphenol A (BPA) 122
Blood measurements 124
Body mass index (BMI) 127
Bottle Materials 92
Bulk sampling 142
Butanal 96
Butylbenzyl Phthalate (BBzP) 129
Butyrate Kinase (BK) 160

C

Capturing larger organisms 39
Carbide method 11
Carbon black pigment 85
Carbon emissions 78
Carbon Nanotubes (CNTs) 110
Carbonyl (CO) 58
Carcinogen 88, 173, 179
Cardiovascular Disease (CVD) 128
Carpenter 12, 31, 95
Casual plastic cutlery 88
Catalase 160
Catalysts stuck 78
Cellular signaling 127
Chemical composition 113
Chemical Interactions 106
Children's hyperactivity 128
Chlorofluorocarbon (CFC) 24
Ciprofloxacin (CIP) 164
Coating materials 113
Codex Alimentarius 89, 90
Codex Alimentarius Commission (CAC) 90
Colorants-Concerns 178
Commercial polyethylenes 12
Competitive wages 6
Confluence 4
Consumer products 122, 123, 129, 143
Copolymers 9, 10, 17, 18, 19, 23, 139, 178
Cosmetic products 94, 102
Cost-effectiveness analyses 145
Cost-effective technology 145
Crystallinity 12, 13, 16, 17, 58, 105, 106, 107, 113, 138, 139
Crystallite formation 13

D

Dairy equipment 122
Deep-sea 32, 72, 98
Deep submarine canyons 99
Degradation processes 57, 58, 104, 105
Di-2-Ethylhexyl Phthalate (DEPH) 20
Dibutyl Phthalate (DEP) 129
Diethylstilbestrol (DES) 123, 176
Differential scanning calorimetry (DSC) 48
Digestive system 68
Digestive tract 40, 61, 73, 78, 114
Diisobutylphthalate (DiBP) 132
Diisononyl Phthalate (DINP) 20
Dimethylformamide (DMFA) 176
Di-n-Butyl Phthalate (DnBP) 132
Di-n-Hexylphthalate (DnHP) 132
Di-n-Octylphthalate (DnOP) 132
Dioxin susceptibility 22
Diurnal solar energy 4
DNA damage 83

Drinking water 94, 95, 97, 142, 143, 145

E

Earth 4, 67, 79, 102
Economic proportionality 38
Ecotoxicological 102
Electronic dipole 50
Electrons and energy 152
Elutriation 42
Endocrine Disruptor Chemicals (EDCs) 81, 85
Endocrine disruptors (EDCs) 86
Endocrine Interference 127
Endocrine system 81, 89, 174, 175
Energy expenditure 60
Energy sources 4
Enhancing crop yields 103
Environmental damage 7, 103
Environmental destruction 5
Environmental evaluation 177
Environmental sustainability 78, 79
Estrogen receptor alpha (Era) 123
Estrogen Receptor beta (ERb) 123
Estrogen-sensitive 125
Ethylene glycol 11, 25
Ethylene Vinyl Alcohol (EVOH) 88, 139
European Commission (EC) 89
European Food and Safety Authority (EFSA) 89
European legislation 110
Expandable Polystyrene (EPS) 132
Extracellular Polymeric Substances (EPS) 146, 154
Extruded Polystyrene (XPS) 132

F

Fewer health risks 178
Fiber spinning 6
Film applications 15
Flame retardant compounds 83
Flame Retardants (FR) 83
Floating debris 55
Floc agglomeration 154
Food and Drug Administration (FDA) 19, 89, 178
Food chain 4, 44, 68, 81, 83, 91, 165, 177
Food chains 68, 69
Food Contact Materials (FCM) 137
Food, Drug and Cosmetic Act 89
Food packaging 19, 87, 88, 96
Forbids plastic microbeads 110
Fourier Transform Infrared Spectroscopy (FT-IR) 73
Freshwater 36
Freshwater organisms 115
Freshwater systems 37, 102, 118

G

Garbage-bag market 14
Garbage disposal 104
Gas chromatography-mass spectrometry (GC-MS) 48
Gas-phase reactors 16, 18
Gelatin 20
Glass fiber 43
Glass transition temperature 82, 96
Global additive 80
Global plastic 46
Global production 6, 12
Groundwater 179
Guideline Values (GL) 90

H

Heavy metals 22, 143, 162, 164, 165, 166, 179
Hetero-aggregation 34
High-Density Polyethylene (HDPE) 9, 47, 57
Higher densities 70

High-molecular-weight propylene polymers 10
High-pressure polyethylenes 15
High temperatures 93
Hormonal system 174
Hormones 63, 81, 114, 126, 128, 174, 175
Human activities 5, 36
Human activity 38, 57
Human consumption 40
Human development 91
Human health 1, 8, 22, 31, 47, 63, 86, 88, 91, 137, 165, 166, 170, 177, 179
Humanity 4
Human serum 122
Human tissue 122
Hydraulic system 27
Hydrology 104
Hydrolytic breakdown 58
Hydrophobic chemicals 80, 81
Hydrophobic contaminants 106, 112, 113

I

Imaging flow cytometry 43
Industrial production 175
Inhaling volatile chemicals 81
Injection-molding machines 27
Inorganic pigments 85
In-vitro Fertilization (IVF) 126
IR spectroscopy 49
Isotactic Polypropylene (iPP) 18

K

Kyoto Protocol 170

L

Laboratory-scale research 154
Leaching of Caprolactam 158
Linear low-density copolymers of ethylene (LLDP) 9
Liquid's surface tension 42
Low-Density Polyethylene (LDPE) 9, 57, 106, 140, 178
Low-molecular-weight liquid polyesters 83

M

Macrolides (MCL) 164
Manufacturing capacity 11, 17
Marine food web 81
Marine snow 72
Marine species 47, 61
Marine waters 33, 36, 37
Market growth 14
Maternal administration 125
Matrix's composition 41
Maximum Contaminant Level Goal (MCLG) 94
Maximum Level (ML) 90
Mechanical stresses 86
Mediterranean Sea 98
Melt Flow Indexes (MFI) 18
Membrane Biofilm Reactor (MBR) 147
Membrane Bioreactor (MBR) 146
Membrane filter systems 144
Membrane technology 148
Metal-containing nanoparticles 115
Metallocene catalysts 9, 13, 16, 17
Metallocene LLDPE resins (mLLDPE) 13
Microbial colonization 31, 57
Microbial communities 55, 59, 146
Microbiological deterioration 46
Microorganisms 34, 59, 64, 140, 146, 148, 149, 150, 152, 153, 154, 155, 158, 160, 161, 166
Microorganisms proliferate 148
Microplastic contamination 33, 45,

48, 72, 108, 143, 145
Microplastics 1, 5, 9, 29, 30, 31, 32, 33, 34, 35, 36, 37, 38, 39, 40, 41, 42, 43, 44, 45, 46, 47, 48, 53, 56, 57, 58, 59, 60, 61, 62, 64, 65, 66, 67, 68, 69, 70, 71, 72, 73, 77, 101, 102, 107, 108, 109, 110, 111, 112, 113, 114, 115, 116, 117, 118, 141, 142, 143, 144, 145, 146, 147, 148, 149, 150, 151, 152, 153, 154, 155, 156, 157, 158, 159, 160, 161, 162, 163
Microscopy 45
Migration 22, 81, 89, 90, 91, 92, 93, 94, 95, 96, 97, 137, 139, 177, 178
Mineral oil 23
Minor portion 17
Molecular weight distribution 13, 16, 58
Molten plastic 79
Moving Bed Bioreactor (MBBR) 147
MP density 119
Munich Plastic Sediment Separator (MPSS) 42
Municipal solid waste 7
Municipal waste 11, 54

N

Nanomaterials 61, 116, 137
Nanoplastics 44, 46, 59, 68, 105, 107, 108, 110, 111, 113, 114, 146, 158, 159, 160, 161
National Toxicology Program's (NTP) 173
Natural geochemical cycles 4
Natural resins 138
Neural function 63
Neuroendocrine circuitry 128
Norwegian Institute for Air Research (NILU) 39
Numerous animal experiments 125

O

Ocean poses 55
Organic contaminants 63, 68, 113, 143, 146, 163, 165, 166
Organic nickel compounds 89
Organometallic complex 16
Organotins 89
Overall migration limit 90

P

Pacific Central Gyre waters 67
Particle behavior 115
Passive diffusion 117
Persistent Organic Pollutants (POPs) 60, 84, 106
Petroleum resources 4
Pharmaceuticals 106, 165
Phenol formaldehyde resin 8
Phthalate plasticizers 81, 82
Phthalates 19, 20, 21, 81, 82, 83, 94, 129, 130, 131, 132, 146, 175
Phthalic anhydride 11
Plastic and microplastic particles 98
Plastic burns 84
Plastic contamination 102, 103
Plastic food packaging 80
Plastic home 83
Plasticizers 82, 135
Plastic pollution 30, 102, 109
Plastic polymers 49, 63, 88, 132, 138
Plastic recycling processes 118
Plastic resins 78, 79, 80
Plastics 3, 5, 6, 7, 30, 54, 55, 58, 59, 62, 64, 78, 79, 80, 81, 94, 98, 102, 103, 117, 118, 121, 122, 143, 158, 162, 169
Plastics business 6, 26

Plastics' manufacturing 6
Plastic waste 31, 34, 46, 55, 56, 57, 59, 81, 82, 98, 118, 142
Polyamides 88, 139
Polybrominated Diphenyl Ethers (PBDEs) 63
Polybutylene Terephthalate (PBT) 135
Polycarbonate (PC) 63, 87
Polychlorinated Dibenzo-P-Dioxins (PCDD) 22
Polycyclic Aromatic Hydrocarbons (PAH) 107
Polycystic Ovarian Syndrome (PCOS) 126
Polyester 25, 44, 93, 136, 150
Polyester film 25
Polyethylene (PE) 56, 86, 110, 135, 138, 163
Polyethylene (PET) 58
Polyethylenes determines 16
Polyethylene terephthalate (PET) 79
Polymer chain 13, 15, 59
Polymerization process 15, 16, 20
Polymer matrix 58, 82, 92, 106, 107, 114, 115, 138
Polyolefin packaging 12
Polyolefins 10, 86, 138
Polyphenols 89
Polypropylene (PP) 57, 86, 110, 135, 138, 145, 156, 163
Polystyrene 10, 11, 12, 23, 24, 25, 44, 63, 79, 83, 96, 102, 110, 132, 138, 146, 163
Polyurethane Foam (PUF) 134
Polyvinyl chloride (PVC) 1, 44, 63, 79
Polyvinyl Chloride (PVC) 9, 94, 110, 139, 163, 170
Possible toxicity 78, 85, 137
Potential consequences 5, 31
Potential sampling approach 72
Prone 19, 58, 60, 112, 138, 143
Prospective epidemiological 126
Protozoa 149, 156
PVC industry 20, 82
PVC pipes 22, 82

Q

Quality Assurance and Quality Control (QA/QC) 65
Quantitative distribution 45

R

Raman spectroscopies 49
Reaction mixture 20
Reactive flame retardant 133, 137
Refrigerators 137
Remotely Operated Vehicles (ROVs) 35
Renal damage 22
Resin granules 23
Resin's glass transition temperature 20
Risk management 109
Rubber chemicals 8

S

Safe Drinking Water Act 94
Saline solution flotation 70
Sample methodology 34
Seawater 43, 48, 72, 106, 164
Separation phase 15
Sequencing Batch Reactor (SBR) 146, 147
Sewage system 144
Sewage treatment 46, 158
Short-chain fatty acids (SCFAs) 161
Skin Damage 171
Sludge 38, 75, 142, 144, 145, 146, 147, 148, 149, 152, 153, 155,

156, 157, 158, 159, 160, 161, 162, 163
Soil biophysical 37
Soils 37
Solar insolation 4
Solid resin 80
Solid waste 7, 12, 118
Standard Operating Protocols (SOPs) 64
Styrene–Acrylonitrile Copolymers (SAN) 23
Styrene–Maleic Anhydride Copolymers (SMA) 23
Submerged Aerated Filters (SAF) 147
Sulfonamides (SDZ) 164
Superoxide Dismutase (SOD) 160
Surface Interactions 112
Surface Microlayer Water (SML) 73
Surface oxidation 117
Sustainable growth 78
Synthetic chemicals 170, 174, 176
Synthetic organic compounds 54, 88
Synthetic particles 103, 143, 155
Synthetic polymers 54, 66, 102, 103, 110, 138, 139

T

Target function 172
Tensile strength changes 104
Terrestrial ecosystems 62, 102, 106
Tetrabromobisphenol A (TBBPA) 133
Tetracyclines (TC) 164
Thermal and thermo-oxidation stabilizers 85
Thermal processing 26, 79
Thermoplastic Elastomers (TPEs) 135
Thermoplastic materials 135
Titanium chloride 15
Tolerable Daily Intake (TDI) 130
Toxic response 117
Transportation 4, 7, 79, 83, 86, 108, 142
Trimethoprim (TMP) 164

U

Ultraviolet (UV) 58, 89
UV-induced deterioration 19

V

Very Low-Density Polyethylene (VLDPE) 13
Vinyl chloride copolymers 9
Vinyl Chloride Monomer (VCM) 19
Vinyl polymers made 8
Virgin plastic resin pellets 72
Visual examination phase 41
Volatile Organic Compounds (VOCs) 170
Volume Of Fugitive Emissions (VOC) 16

W

Wastewater 33, 35, 56, 60, 75, 104, 134, 142, 144, 145, 146, 147, 148, 149, 152, 153, 155, 157, 159, 161, 166
Waste Water Treatment Plants (WWTPs) 103
Water phase 157
Water-resistant polymers 88
Weathering processes 1, 113, 116, 117
Whether aquatic 40

Z

Ziegler-Natta catalysts 9, 17